A BIRDWATCHERS' GUIDE TO TRINIDAD & TOBAGO

William L. Murphy

BIRD WATCHERS' GUIDES

Prion Ltd.
Cley

ACKNOWLEDGEMENTS

This book is dedicated to Donald H. Messersmith, a superb educator at the University of Maryland and a dear friend to whom I owe my love of the birds of Trinidad and Tobago. By instructing me in ornithology and entomology, training me as a bird tour guide, and later hiring me as tour escort on my first trip to Trinidad and Tobago, he enabled me to experience a tropical blend of birdlife that has since become an obsession with me and resulted in many return visits to the islands – and in the production of this guide.

Among the many other people I would like to thank for their inspiration and help are my parents, Bernadette and Lawrence Murphy; Kenny Calderon; Peter Cox; Davis Finch; Richard ffrench; Newton George; Rob Gibbs; Cory Gildersleeve; Geoffrey Gomes; Floyd Hayes; Adolphus James; Martyn Kenefick; David Ramlal; Jogie Ramlal; Rudal Ramlal; Gerard and Oda Ramsawak; Courtenay Rooks; David Rooks; April Sterling; and Graham White.

I also thank Margaret Schaeffer, president and owner of Caligo Ventures, and her sister, Denise Rasulo, for providing information on the Asa Wright Nature Centre and in particular for their dedication to getting naturalists to and from Trinidad and Tobago over the years; to Ryan D'Arcy, Marketing Officer of the Tourism & Industrial Development Company of Trinidad & Tobago Limited, for verifying myriad details relating to commercial enterprises on both islands; to Richard S. Shirley for permission to use his image of Mahase Ramlal; to the late Don R. Eckelberry for perpetual rights to his illustrations in the book *A Naturalist in Trinidad*, by C. Brooke Worth; and to the late Jeanne Gildersleeve for making the first incarnation of this book aesthetically pleasing.

Finally, I wish to express heartfelt thanks to my former editor and friend, the late Sandra Sherman, for her patience, expertise, and encouragement.

CONTENTS

Golden-headed Manakin

INTRODUCTION

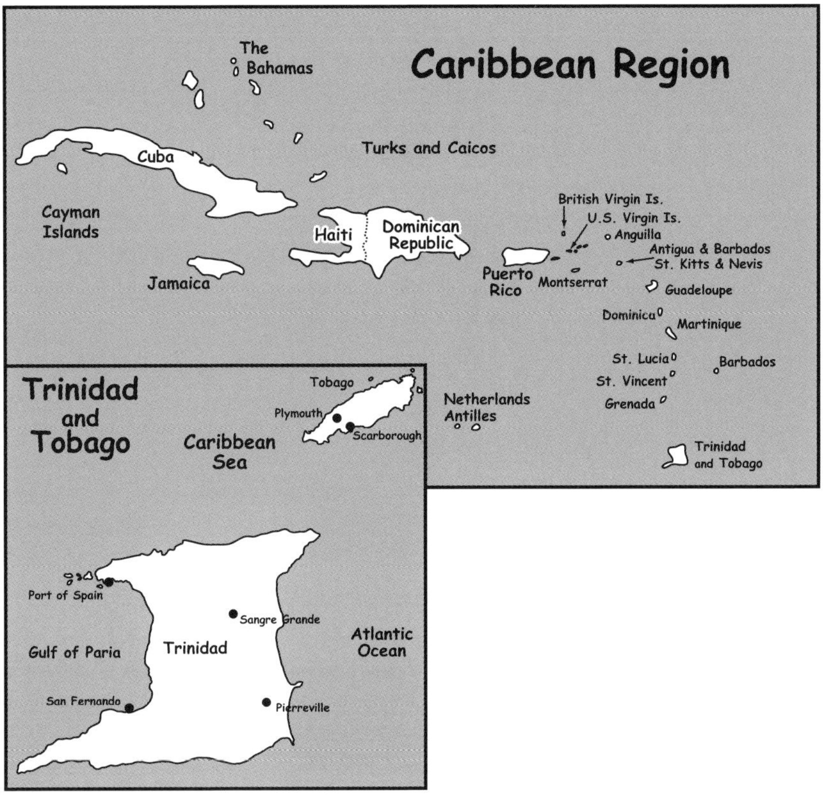

How to Use This Book

This guide provides information for birders about Trinidad and Tobago and the birds found there. The location sections give details about birding in specific areas, while the specialities section gives details about individual species. Bar graphs provide visual information concerning the distribution of species throughout the year.

Do familiarize yourself with the geography of the islands and carry a road map. Distances on the maps are shown in kilometres. Unless otherwise stated, these distances represent the distance from the last place mentioned, not from the starting point.

Except for some of the seabirds, waders, gulls and terns, the majority of species known to have occurred on either island are admirably described and illustrated in Richard ffrench's book, *A Guide to the Birds of Trinidad and Tobago*, which should be used in conjunction with this book.

Nearly 400 species have been recorded on Trinidad, partly as a result of the great diversity of habitats, which range from ocean to mountains and from scrub-covered deserts to rainforests. The diversity of habitats include elfin forest, montane rainforest, lowland rainforest, savannah grassland, freshwater swamp, saltwater swamp, freshwater reservoir, ocean beach, and open ocean, as well as a wide variety of cultivated areas and formerly cultivated areas that are regenerating naturally. In addition, an abundance of perennially available fruits and seeds provides forage for resident species and migrants from both North and South America.

History

The human saga on Trinidad and Tobago began some 7,000 years ago, but the written history dates back only to the discovery of the islands by Christopher Columbus on 31 July 1498. He named the island "La Trinidad" in honor of the Blessed Trinity. Two Amerindian tribes, the Arawaks from the Orinoco region and the Caribs from the Amazon region, occupied Trinidad, and the Caribs occupied Tobago.

The Amerindians were badly mistreated by the Spanish, and diseases such as smallpox claimed huge numbers of lives. After the Spaniards, the French next colonised the islands, and more than one million African slaves were imported. In 1797 the British conquered Trinidad, which remained a British colony until 1962.

On 1 August 1834 Great Britain abolished slavery. East Indians were induced to emigrate to Trinidad, where they prospered. Today East Indians account for about 45% of the total population, with Afro-Americans accounting for another 45% and Chinese, mostly business owners, accounting for about 5%.

Tobago became extremely wealthy during this period because of its production of sugar cane. When Tobago's sugar cane-based economy collapsed in 1884, Trinidad formally annexed Tobago. In the mid-1800s oil was discovered in Trinidad. This discovery permanently changed the economy of the islands, allowing them to export vast quantities of petroleum and petroleum products worldwide.

In 1956 a political party called the People's National Movement (PNM) was formed, with Dr. Eric Williams as its head. The PNM became the first Party government in Trinidad and Tobago. Partly as a result of its nationalistic efforts, Trinidad and Tobago attained independence on 31 August 1962. On 1 August 1976, Trinidad and Tobago was declared a Republic but remains a member of the British Commonwealth.

Environment

Trinidad and Tobago are the southernmost pair of islands in the Caribbean chain, small chips of South America that lie just off the coast of Venezuela, only about 10° north of the Equator. The southern coast of Trinidad lies just 13 km (8 miles) north of the South American mainland, across the Columbus Channel from Venezuela. Tobago lies 34 km (21 miles) northeast of Trinidad.

Temperatures average 29°C (84°F) during the day and 23°C (74°F) at night. The national language is English. The population is 1,262,366 (July 2003 estimate), of which 54,084 reside in Tobago. Electricity is 110 and 220 volts, 60 cycle. Trinidad and Tobago are four hours behind Greenwich Mean Time.

Biologically and geologically the islands of Trinidad and Tobago are related only distantly. Trinidad was formed in the same series of geological events that shaped the rest of the South American mainland. To that initial body the Orinoco River contributed considerable sediment. The formation of the Northern Range finished the picture.

This mountain range consisted of scrapings from a tectonic plate being subsumed beneath the one on which Trinidad rests. This process continues today and is the source of the seismic activity in the area as well as such phenomena as geothermal vents and mud volcanoes.

Trinidad became an island about 10,000 years ago at the end of the Wisconsinan glaciation of the Pleistocene Epoch, during which much of the earth's seawater was held in the ice caps. As the ice melted, the sea submerged all but the highest peaks of the land bridge that connected Trinidad with Venezuela. The flora and fauna of Trinidad thus became isolated from that of the mainland.

In contrast, Tobago rests on an eastward-moving tectonic plate that once was adjacent to the Lake Maracaibo region of western Venezuela. Even today the flora, fauna, and geology of Tobago are much more closely allied with those of western Venezuela than they are with those of Trinidad.

Geography

Trinidad

Trinidad is somewhat rectangular in shape, about 80 × 60 km (50 × 37 miles), with westward-pointing extensions in the northwest and southwest. Trinidad covers an area of about 4,830 km^2 (1,865 $miles^2$) and is relatively level except for three east-west mountain ranges.

The Northern Range extends the entire width of Trinidad, from the Chaguaramas Peninsula in the west to Galera Point in the east. This mountain range is actually the easternmost spur of the Andes. It averages the highest of the three ranges, with El Cerro del Aripo and El Tucuche rising to over 940 m (3,084 ft.). Much of the Northern Range is covered with lush rainforest and coffee-cocoa-citrus plantations, both active and abandoned. The eastern third of the Northern Range is very sparsely populated, contains pristine rainforest, and receives more rainfall than any other part of Trinidad, as humid air from the Atlantic rises over the peaks, cools, and releases its moisture. In the wettest areas rainfall may reach 500 cm (200 in.) per year but elsewhere rainfall rarely exceeds 200 cm (80 in.) per year. The Gulf of Paria, a splendid natural harbour, separates Trinidad from Venezuela to the west. A chain of five arid islands – Gasparee, Monos, Huevos, Chacachacare, and Patos (Venezuelan) – dots the 19 km (11.8 mile) channel ("The Dragon's Mouth") that lies between the western tip of northern Trinidad and the eastern tip of the Paria Peninsula of Venezuela.

Trinidad's Central Range stretches diagonally from the Gulf of Paria in the west across central Trinidad and northeastward to the Atlantic Ocean, reaching a maximum elevation of 307 m (1,007 ft.) on Mount Tamana. To the south are forests, swamps, private oil fields, and the bituminous Pitch Lake.

In the southeastern corner of the island are the low (250 m/820 ft.) Trinity Hills. The Columbus Channel, 13 km (8 miles) wide, separates southern Trinidad from Venezuela.

The remaining areas of Trinidad are mainly lowlands that are or were almost entirely cultivated except for the North Oropuche and Nariva Swamps in the east and the Caroni, South Oropuche, and Icacos Swamps in the west and southwest. Trinidad's shorelines range from palm-fringed sandy beaches to vertical rock cliffs.

Geography

Tobago

Tobago is shaped like a northeastward-pointing fish, about 40 km long and 12 km at its widest (25 × 7.5 miles), with an area of 300 km^2 (116 $miles^2$).

Rainforest covers most of the rugged spine of mountains called the Main Ridge that extends almost two-thirds of the length of the island. Tobago's overall topography is more rugged than that of Trinidad but the mountains are not as high; the highest point, Pigeon Peak, reaches only 576 m (1,890 ft.). Partly because of a lack of high peaks or ridges to deflect moist breezes upwards, Tobago is drier overall than Trinidad.

The southwestern area of Tobago consists of highly fertile flatlands that overlie coral/limestone. This area is home to the majority of Tobagonians. Here, too, are found the only extensive wetlands in Tobago. The coastline of Tobago is diverse, being sandy in some stretches and rocky in others. Offshore islands, principally St. Giles and Little Tobago Island, both of which lie off Tobago's northern coast, harbour nesting seabirds.

Natural History

The fauna and flora of Trinidad and Tobago resembles that of nearby Venezuela partly as a result of the islands' continental origin and their proximity to South America. Moreover, Trinidad is unique as the link between Amazonian and West Indian biota. The list of species is impressive, with records of 99 species of mammals, 96 reptiles, 37 amphibians, and 617 butterflies. Similar diversity is found in other animal groups and among the plants, many of which are endemic to Trinidad or Tobago. As of 30 June 2004, 457 species of birds had been recorded, of which about 250 are known to breed. Trinidad and Tobago offers a fine introduction to birding in the Neotropics without being overwhelming. No other area in the West Indies – indeed few areas of comparable size elsewhere in tropical America – can match the diversity of species. Migrant species of North American origin can be observed on both islands from October until April, and migrant species of South American origin can be observed on both islands from June until August.

During a visit to Trinidad it is possible for birders to record 140-180 species depending on the time of year, with another 20-30 species possible with a visit to Tobago. Persons venturing no farther afield than the veranda ('gallery') of one of the main guesthouses routinely compile lists of 40-60 species in one day.

PRE-TOUR INFORMATION

Immigration and Customs

Entering and leaving Trinidad and Tobago is relatively painless. An immigration card must be completed during the incoming flight and must be surrendered on departure. Visitors are allowed to import 50 cigars, 200 cigarettes or one pound of tobacco, and one quart of distilled spirits. A payment of TT$100 is required on departure for airport and security taxes.

Every visitor must possess a valid passport. Although no formal agreement exists, citizens of Venezuela or the U.S. who are seeking entry exclusively for a holiday do not require visas for stays of two weeks or less (Venezuelans) or two months or less ((U.S citizens). UK citizens do not require visas for stays of less than three months.

Entry visas are not required for citizens of British Commonwealth countries, except for Australia, India, New Zealand, Nigeria, Papua New Guinea, Sri Lanka, Tanzania, and Uganda. Pakistan is treated as a Commonwealth country for visa purposes. Citizens of the following countries must obtain visas prior to entry into Trinidad and Tobago: Albania, Bulgaria, Cuba, all of the former Czechoslovakian states, Hungary, North Korea, the Peoples' Republic of China, Poland, the Republic of China (Taiwan), Yugoslavia, Romania, all of the states of the former USSR, and Vietnam.

Money

Currency is the Trinidad and Tobago dollar (TT dollar or TT$, pronounced "tee-tee"), which has a floating exchange rate. At the time of writing (2004) the exchange rate was TT$6.25 to the US dollar and TT$11.00 to the UK pound. When spending foreign currency in Trinidad and Tobago, be sure to insist on the full exchange rate at hotels and shops; cashiers will often use a lower rate.

Beware of black-market dealers at Piarco Airport who may offer one and a half times the current exchange rate. Local banks will not exchange unspent TT$ for other currencies unless a bank receipt can be produced for at least the amount to be redeemed. The 'cambio' at Piarco Airport offers a competitive exchange rate and provides receipts. Guides and hotel managers often have a genuine need for foreign currency and will provide a fair exchange rate for small sums. The preferred currency throughout the islands is the US dollar.

Inoculations

No inoculations, vaccinations, or anti-malaria pills are required for a visit Trinidad and Tobago, but a yellow fever vaccination may be advisable since some remote forests in southeastern Trinidad are inhabited by Red Howler Monkeys that harbour the yellow fever virus.

TRAVEL INFORMATION

How to Get There

British West Indian Airways (BWIA), the national airline of Trinidad and Tobago, has been flying for 63 years. BWIA usually offers reasonable prices for trips to Trinidad and Tobago.

In Europe BWIA serves Heathrow International Airport. In the U.S. BWIA serves Kennedy International Airport in New York, Dulles International Airport near Washington, D.C., and Miami International Airport in Florida. In Canada BWIA serves Pearson International Airport in Toronto.

All BWIA flights to Trinidad land at Piarco International Airport in north-central Trinidad, with linking flights to Crown Point Airport in Tobago. Many flights stop once, some in Barbados, St. Lucia, or Antigua, and others in Aruba. The flight from Miami lasts about four hours, from New York five to six hours, and from London seven to eight hours. BWIA's website is http://www.bwee.com

Regarding other carriers, Air Canada serves Trinidad from Toronto and American Airlines serves Trinidad from Miami twice daily. Condor (a German charter service) and British Airways both offer flights directly to Tobago. Other charters (Air Transat, Canada 3000, World Airways, etc.) continue to serve Trinidad and Tobago from most major cities.

Airports

Trinidad: Piarco International Airport is centrally located about 25 km (15.5 miles) east of Port-of-Spain. Piarco boasts a 3,200 m (10,500 ft.), runway and the newest terminal in the Caribbean, which is open 24 hours a day, seven days a week.

Tobago: Crown Point International Airport is the only airport on the island of Tobago and is very small. Located at the extreme south-western point (Crown Point), it is within walking distance of beautiful beaches and several of the island's hotels.

For detailed information about either airport, please visit http://www.tntairports.com/

Travelling in Trinidad and Tobago

Travelling in Trinidad and Tobago is easy. Taxis are found everywhere except in the most remote locations, even late at night. Buses adhere to regular schedules along all highways and offer extremely reasonable fares.

All licence plates in Trinidad and Tobago begin with one of four letters: H (for hire), T (truck), P (personal vehicle), or R (rental). It is possible to hail a taxi displaying an H plate from the roadside. Taxi fares are somewhat higher than bus fares, but taxis are much more common than buses and the fares are reasonable. H-plate vehicles sometimes travel along specific routes and therefore may not deliver passengers to precise destinations. Drivers often drop people off at major intersections, where another taxi can be secured for the rest of the journey. Alternatively it may be possible to negotiate with a driver to take you to your destination for an additional fee. Officially designated

Maxi-taxi

white high-topped minibuses (maxi-taxis) are fairly easy to find and also follow regular routes, stopping wherever potential passengers flag them down. Fares are somewhat lower for maxi-taxis than other taxis. Hotel managers are happy to recommend drivers and will even engage one if requested.

Driving

In Trinidad and Tobago people drive on the left. Note the white mileposts along many of the roads. The bars under the mile numbers represent the number of quarter miles from the last full mile marker. Vehicles often stop in the road, even in heavy traffic, while the driver chats with friends on the roadside – this is called 'liming'. A word of warning: it is not wise to pick up hitchhikers.

Visitors in possession of valid driving licences issued in the UK, USA, Canada, France, Germany, or the Bahamas may drive legally in Trinidad and Tobago for up to three months. International licences are also accepted.

Petrol stations are closed for most or all of the day on the following public holidays: New Year's Day (1 January), Carnival (the Monday and Tuesday before Ash Wednesday), Good Friday (the Friday preceding Easter), Easter Monday (the day after Easter), Whit Monday (movable), Corpus Christi (movable), Labour Day (movable), Emancipation Day (1 August), Independence Day (31 August), Republic Day (24 September), Christmas Day (25 December), Boxing Day (26 December), Eid Ul Fitr (movable), and Festival Divali (movable).

Airports are usually the best places to rent a car. At the time of writing, a small car such as a Nissan Sentra with automatic transmission and air conditioning will cost TT$56 per day including insurance, tax, and VAT. Prices do not vary much but some companies require a TT$3,000 deposit or a credit card for security. Nearly all car-rental companies give the seventh day free; payment is required only to cover insurance for that day. To rent a car, drivers must be at least 25 years of age, possess a valid driving licence, and have a minimum of three years' driving experience.

Renting a Car

The following car-rental companies are located at Piarco International Airport or near Crown Point International Airport. They are friendly, fairly priced, and reliable.

In Trinidad:

Auto Rentals (AR)	(868) 669-2277 ('669-CARS')
Econo-Car Rentals e-mail: econocar@trinidad.net	(868) 669-2342
Johnney Car Rentals	(868) 669-2226
Kalloo's Car Rental	(868) 669-5673
Singh's Auto Rentals Ltd. e-mail: singhs1@tstt.net.tt	(868) 669-5417
Sue's Auto Rentals e-mail: suesauto@tstt.net.tt	(868) 669-1635
Thrifty Car Rental e-mail: eontab@tstt.net.tt	(868) 669-0602

In Tobago:

Auto Rentals (AR)	(868) 639-0644
Carlton James Car Rental	(868) 639-8084
Econo-Car Rentals	(868) 639-8378
Peter Gremli's Car Rental	(868) 639-8400
Rattan's Car Rental	(868) 639-8271
Quashies's Car Rentals	(868) 639-8427
Singh's Auto Rentals Ltd.	(868) 639-0191
Thrifty Car Rental	(868) 669-8507
Toyota Rent-a-Car	(868) 639-7491

Trinidad–Tobago Ferry

A voyage aboard the 700-passenger ferry *M. F. Panorama* or the 820-passenger ferry *M. F. Beaufort* as they travel between the two islands could be considered a mini pelagic trip. Birds seen from the ferry include shearwaters, petrels, tropicbirds, boobies, terns, gulls, and possibly jaegers and skuas. A Black-browed Albatross was observed far to the northwest of Tobago a few years ago, so there is no reason to believe that an albatross might not occasionally stray into Trinidad and Tobago waters.

The domestic ferry service is operated by the T&T Port Authority. It provides transportation for passengers, light vehicles, and cargo. Onboard services include dining and bars.

Tickets, valid for 90 days from date of purchase, can be obtained from the Port-of-Spain and Scarborough offices of the Port Authority as well as at Royal Bank branches in Arima, Chaguanas, San Fernando, and Point Fortin. Office hours at the Port-of-Spain office are Monday to Friday, 7.30 a.m. to 4 p.m. Passenger ticket sales close two hours before departure. Vehicles must be on board two hours before sailing.

Sailing times: Monday to Friday: departs Port-of-Spain, Trinidad at 2 p.m. and departs Scarborough, Tobago at 11 p.m.
Saturday: departs Port-of-Spain at 11 a.m. and departs Scarborough at 11 a.m.
Sundays and Public Holidays: departs Port-of-Spain at 11 a.m. and departs Scarborough at 11 p.m.

Special sailings are scheduled during peak periods and public holidays. The voyage takes approximately five-and-a-half hours. For current information visit http://www.tntisland.com/ferry.html/

Fares (subject to change):

Passengers	
Economy class	TT$50 round trip
Tourist class (reclining seats)	TT$60 round trip
Cabin (double occupancy)	TT$160 round trip
Children under 12	half price
Infants under 3 years	free
Light vehicles	TT$600 round trip

For more information:

Port-of-Spain	(868) 625-4906/3055
Tobago	(868) 639-2417 (passengers) (868) 639-2181 (cargo)

ACCOMMODATION

Only hotels and guest houses popular with birdwatchers are listed below. The Asa Wright Nature Centre and the Pax Guest House Mount St. Benedict are both top-notch places to stay. Both welcome those travelling independently.

Trinidad

Asa Wright Nature Centre

Verandah at the Asa Wright Nature Centre

The Asa Wright Nature Centre, the former Springhill Estate, is located in the Northern Range at an elevation of 360 m (1,181 ft.), 12 km (7.5 miles) north of the city of Arima ('ah-REE-muh'). Formerly a cocoa-coffee-citrus plantation, it is now a wildlife sanctuary covered mostly by reclaimed secondary forest that is largely surrounded by impressive climax rainforest. Several kilometres of trails offer spectacular birding opportunities. Meals at the Centre include many delicious local specialities, served buffet style at set times. Excellent picnic lunches are provided for field trips. Traditional afternoon tea is served, often enlivened by the anvil-like chorus of Bearded Bellbirds from the valley below. Before dinner, guests enjoy a complimentary rum punch served on the veranda. A true retreat, no telephone is available at the Centre but public phones are available in Arima, a 15-minute drive down the mountain. The Centre maintains constant contact with its Arima office by means of short-wave radio.

Electricity at the Centre is 120 volt, 60 cycle alternating current. Some rooms are in the estate house and some are in separate bungalows. All have en suite bathrooms with a shower.

The Manager
Asa Wright Nature Centre and Lodge
P.O. Box 4710
Arima, Trinidad
(868) 667-4655 voice
(868) 667-0493 fax
e-mail: asaright@tstt.net.tt
web-site: http://www.asaswright.org

or

Caligo Ventures Inc.
156 Bedford Road
Armonk, New York 10504, USA
(800) 426-7781 from the U.S. and Canada
(914) 273-6333 from outside the U.S.

Pax Guest House Mount St. Benedict

At the edge of the Northern Range, with the ornate tapestry of the blue-green forest on Mount Tabor at its back and the lush Plains of Caroni at its feet, lies the Pax Guest House Mount St. Benedict. Centrally located in the cool heights (250 m/820 ft.) of the largest and oldest (1912) Benedictine monastery in the Caribbean, Pax boasts 19 rooms, three tennis courts, and several nature trails on a 243 ha. (600 acre) estate. The area contains two distinctly different types of evergreen forest: softwood (Caribbean pine) and hardwood. Guided tours through the estate are available. Pax is only five minutes from the campus of the University of the West Indies (UWI; pronounced 'YOU-ee'). The proximity of Pax to UWI, combined with the availability of basic laboratory and conference facilities, makes Pax a perfect base for visiting scientists.

Pax is centrally located for excursions in all directions and is surrounded by bird-rich habitat. The manager, Gerard Ramsawak, is a perfect host, providing delectable Caribbean food to be enjoyed in the pleasant ambiance. Piarco Airport is about 11 km (7 miles) to the east and Port-of-Spain is 12 km (7.5 miles) to the west.

Electricity at Pax is 120 volt, 60 cycle alternating current. Most rooms are on the second floor and bathrooms are mostly shared, with separate men's and ladies' toilet facilities. There are also self-contained (efficiency) rooms. Breakfast is served from 7.30 to 9 a.m., lunch at 12.30 p.m., and dinner at 7.30 p.m. Birders can arrange to have breakfast served as early as 5.30 a.m. or to have fruit and beverages set out for them the night before. Enormous lunches are packed every day for picnics in the field.

Pax Guest House Mount St. Benedict
Mount St. Benedict
Tunapuna, Trinidad
(868) 662-4084 voice and fax
(868) 645-4232 voice and fax
e-mail: pax-g-h@trinidad.net
web-site: http://www.paxguesthouse.com/

Jasmine's Ocean Resort

Surrounded by lush evergreen rainforest, this tropical nature reserve is located on the northeastern coast of Trinidad on a secluded cove at Balandra. The resort offers 14 double rooms ranging from standard to deluxe to superior, along with a restaurant and pool. Along the forest trails it is possible to birdwatch and enjoy panoramic views of the northeastern coast.

Jasmine's Ocean Resort
#1 Simmons Drive
Rampanalgas Village, Toco Main Road
Balandra, Trinidad
(868) 670-4567 voice
(868) 670-0992 fax
e-mail: info@jasmines.net
web-site: http://www.jasminesresort.com/

Laguna Mar Nature Lodge & Cocos Hut Restaurant

Laguna Mar is an oceanfront retreat for naturalists. Located on the north coast of Trinidad on the outskirts of the old French fishing village of Blanchisseuse, the property extends from the sandy beaches of Blanchisseuse Bay up into the mountains of the Northern Range. The lush tropical rainforest is interspersed with cultivated crops such as coconut, cocoa, coffee, citrus, and other fruits. Meandering through the valleys and fed by numerous freshwater springs, the Marianne River flows into the ocean at the northeastern end of the property. At its mouth is a beautiful lagoon that is perfect for swimming. During nesting season giant Leatherback Turtles come ashore and lay their eggs on the soft sandy beach.

The Laguna Mar Lodge has six guest rooms, all with en suite bathrooms and wide balconies cooled by constant ocean breezes. The nearby Cocos Hut Restaurant, which occupies a 71-year-old agricultural drying and storage hut, is open seven days a week and serves local specialities, especially seafood. Much of the food is grown on the estate.

Laguna Mar Beach Resort & Nature Lodge
Mile Marker 65 1/2
Paria Main Road
Blanchisseuse, Trinidad
(868) 628-3731 voice
(868) 628-3737 fax
e-mail: info@lagunamar.com
web-site: http://www.lagunamar.com/

Le Grande Almandier

This guesthouse, located on the beach at Grande Riviere in northeastern Trinidad, was built by community members Cyril and Wendy James. It consists of five rooms that sleep up to four people each and one room that sleeps up to five. The restaurant serves creole dishes, many with fish caught less than two hours beforehand. Cyril will also take guests deep-sea fishing to catch the night's dinner.

e-mail: info@legrandealmandier.com
web-site: http://www.legrandalmandier.com/

Mt. Plaisir Hotel

This 9-room hotel is located on secluded Grand Riviere Bay in northeastern Trinidad. The Caribbean is at the front, it backs onto a majestic tropical rainforest, and a freshwater stream runs into the ocean on either side. Mt. Plaisir is a superb base for birding in northeastern Trinidad, especially to see the rare Trinidad Piping-Guan. During the nesting season, Leatherback Turtles lay their eggs in the sand just outside the rooms. Later in the season turtle hatchlings can be seen appearing in the sand and making their journey to the sea. Many ingredients used in the preparation of the excellent food served at Mt. Plaisir are grown on the estate, and delicious breads and pizzas are prepared daily.

Mt. Plaisir Estate
Grand Riviere, Trinidad
(868) 670-8381 voice

(868) 670-0057 fax
e-mail: info@mtplaisir.com
web-site: http://www.mtplaisir.com/

Surf's Country Inn

This popular seafood restaurant in the North Coast seaside village of Blanchisseuse offers six guest rooms in adjoining buildings. Surf's is located on a wooded hillside overlooking the Caribbean. The Northern Range is behind the inn, with excellent birdwatching opportunities along the nature trails and rivers.

Surf's Country Inn
LP 197, Lower Village
Blanchisseuse, Trinidad
(868) 669-2475 voice
(868) 669-3016 fax

Windrush Bed and Breakfast

Eminel Cooper owns this private bed and breakfast. Located not far from Surf's Country Inn in the seaside village of Blanchisseuse, it is a very pleasant, cool, and extremely neat house with a beautiful veranda overlooking the Blanchisseuse coastline. Nearby are several sandy beaches, and the nature trails of the Northern Range are not far away. Other attractions in the area include the Marianne River, Paria Bay and waterfall, forests of Mora trees, and several former cocoa and coffee estates.

Cooper also manages a small, very simple two-bedroom bungalow in Blanchisseuse, just across the road from Marianne Beach. Each bedroom has one double and one single bed, a bathroom with a cold water shower, living/dining area, a patio off the living room, and a kitchen equipped with the necessary utensils. The constant sea breeze makes air-conditioning unnecessary.

Windrush
Paradise Hill, Upper Village
Blanchisseuse, Tobago
(868) 664-6114 voice
e-mail: delisle@carib-link.net
web-site: http://members.tripod.com/windrushtt

Tobago

Arnos Vale Hotel

Nestled against the gentle slopes of the 162 ha. (400 acre) Arnos Vale Estate, the Arnos Vale Hotel is a tropical paradise of exotic blossoms, beautiful birds, a secluded beach, and a pristine coral reef. Located about 15 km (9 miles) from Crown Point Airport, this 30-room hotel has the charm and grace of the bygone era when the estate was a sugar plantation while offering the facilities and comfort of an upmarket resort. The local birds are fed daily at a 5 p.m. tea ceremony. Guests are able to hand-feed them.

The facilities are located on a hill that slopes gently down to the enchantingly private Arnos Vale Bay. Rooms can be reserved in the estate house, a beachside facility, or hillside bungalows. A nature trail on the premises allows birdwatchers easy access to several different kinds of habitat.

Arnos Vale Hotel
P.O. Box 208
Plymouth, Tobago
(868) 639-2881/2882 voice
(868) 639-4629 fax
e-mail: reservations@arnosvalehotel.com
web-site: http://www.arnosvalehotel.com/

Blue Waters Inn

At the northeastern tip of Tobago and certainly in one of the most secluded parts of the island lies the Blue Waters Inn. Located near the village of Speyside, bordering Batteaux Bay and facing Little Tobago Island, the Blue Waters Inn offers 31 rooms, three self-catering apartments, and three bungalows. The hotel has an open-air restaurant, a bar, and a dive shop on the premises. Several species of birds fly through the dining room, and Ruddy Turnstones often pick up crumbs from under the tables. The area abounds with birdlife and is only a short drive from the Roxborough-Bloody Bay Road, the route to Tobago's virgin rainforest. Daily boat trips depart from the Blue Waters Inn to Little Tobago Island.

Blue Waters Inn
Speyside, Tobago
(868) 660-4077/4341 voice
(868) 660-5195 fax
e-mail: bwi@bluewatersinn.com
web-site: http://www.bluewatersinn.com/

Cocrico Inn

The Cocrico Inn is a simple, inexpensive, family-run hotel about 10 km (6.2 miles) from the airport, just south of the Arnos Vale Hotel. It is located in the town of Plymouth, a typical Tobagonian fishing community, where the visitor has an opportunity to converse with 'real' Tobagonians. The managers, Nick and Bev Sanford, are from the US and know the type of amenities birders like. All 16 rooms have an en suite bath and shower. Early breakfasts are no problem. The villa is a good place to stay because of its natural setting.

Cocrico Inn
North & Commissioner Streets
P. O. Box 287
Plymouth, Tobago
(868) 639-2961 voice
(868) 639-6565 fax
e-mail: hewlette@tstt.net.tt
web-site: http://www.hews-tours.com/cocricoinn.html

Cuffie River Nature Retreat

Cradled in the beautiful Runnemede Valley, perched above gently flowing streams and mountain pools at the centre of Tobago's North Coast and on the edge of Tobago's virgin rainforest, lies the Cuffie River Nature Retreat.

Originally a cocoa estate, the property has been reclaimed by secondary forest and is now covered with lush bamboo groves, wild heliconias, and cocoa trees. The area is a paradise for birders and nature lovers. Guided walks can be arranged at the lodge. The 10 guest rooms are spacious, tastefully furnished, and have private balconies and baths. A variety of native meals are home-cooked throughout the day.

Cuffie River Nature Retreat
Runnemede, Tobago
(868) 660-0505 voice
(868) 660-0606 fax
e-mail: info@cuffie-river.com
web-site: http://www.cuffie-river.com/

Footprints Eco Resort

Located 35 minutes from Crown Point on the Caribbean side of Tobago, this nature reserve is being extensively reforested with native species. There are numerous nature trails suitable for birding. The hotel grows fresh organic herbs and vegetables for the resort's restaurant. A saltwater pool is drawn directly from Culloden Bay, which supports a reef suitable for snorkelling. The rooms and villas are constructed of locally grown woods including recycled wallaba and teak reclaimed from waste wood. The roofs are made from the leaves of the timit palm, a durable and attractive covering that served as the first material used for roofing in Trinidad and Tobago. The accommodation is unobtrusive and designed to enhance rather than overshadow the rolling green hills and to blend harmoniously with the ridge tops and valleys. All rooms and villas have wonderful views of the Caribbean.

Footprints Eco Resort
Cullodan Bay, Tobago
(868) 660-0118 voice
(868) 657-6991 fax
e-mail: info@footprintseco-resort.com
web-site: http://www.footprintseco-resort.com/

Grafton Great House

Grafton Great House, on Grafton Estate, is located on the leeward (Caribbean) side of Tobago near the village of Black Rock. This 1930s plantation house was home to Mrs Eleanor Alefounder. According to the present manager (Margretta McWilliam of Cumbria, England), Mrs Alefounder had a lifelong interest in wildlife and conservation. After Hurricane Flora devastated Tobago in 1963, destroying the birds' habitat and natural food, she helped feed the birds brought to her estate house by starvation, including those species normally too shy to break cover. Mrs Alefounder continued to feed the birds daily on her 81 ha. (200 acre) estate, particularly in the immediate vicinity of the house. The proximity of so many birds fearlessly coming to her veranda attracted visitors from all over the world. Many species also flew through the house. Bananaquits were everywhere, Rufous-vented Chachalacas and Blue-crowned Motmots could be hand-fed, and the motmots showed a decided preference for cheese.

Eleanor Alefounder died in 1983 but her work is now carried out by Trustees. The copra shed near the house has been restored and converted into an observation centre where Rufous-vented Chachalacas and Blue-crowned Motmots can be seen. The former wagon trails through the copra plantation are kept reasonably clear of brush and have become nature trails. The major portion of the estate was willed by Mrs Alefounder as a bird and wildlife sanctuary for all time, while she left the house to her family. Grafton Great House is essentially a beloved family home, but between family visits it is rented to ornithologists and others who appreciate peace and tranquility. The house has been renovated to retain its original appearance.

The entire house can be rented. It has three large bedrooms, with an additional small bedroom for a child. The living and dining areas are spacious, with polished wooden floors. The kitchen is large and fully equipped. For a small additional charge a maid can be hired to cook and clean. Modern laundry facilities are also available. The trails on the grounds are clearly marked and lead to areas inhabited by such Tobago specialties as Blue-backed Manakin and Olivaceous Woodcreeper. Enormous trees near the great house are festooned with night-blooming *Cereus* cactus. The elevated site of the great house takes advantage of cooling sea breezes and affords magnificent views of Stonehaven and Courland Bays and the villages of Black Rock and Plymouth. The garden is secluded, with terraced lawns and an orchard. The beach is about 0.3 km (900 ft.) away and is easily accessible by foot or car.

Grafton Great House
Black Rock, Tobago
022-971-6568/6721 voice and fax
(this telephone number is in the U.K.)

Kariwak Village

The Kariwak Village is a delightful, intimate 24-room hideaway within walking distance of Crown Point Airport. In addition to serving some of the best food in Tobago and providing early breakfasts and packed lunches for birders, the Kariwak offers a swimming pool, morning stretch classes, tai chi, qi gong, yoga, and live music on Friday and Saturday nights. Over 100 species of tropical plants can be found on the property, along with 20 species of birds. The Kariwak is convenient for Crown Point, Pigeon Point, Buccoo, and Grafton.

Kariwak Village
P.O. Box 27
Scarborough, Tobago
(868) 639-8442 voice
(868) 639-8441 fax
e-mail: kariwak@tstt.net.tt
web-site: www.kariwak.co.tt

Palm Tree Village

One of the newest facilities in Tobago, the Palm Tree Village resort features both independent villas (two or four bedrooms) and 20 luxury twin hotel rooms along with a restaurant, swimming pool, bar, conference rooms, tennis court, and horseback riding. The octagonal villas have fully equipped kitchenettes, air-conditioned bedrooms, satellite television and movie channels, spacious living rooms, and ample patio space. The Palm Tree Village is located at the southeastern tip of Tobago on the windward (Atlantic) side. Activities available include dancing, golf, tennis, swimming, snorkelling, scuba diving, waterskiing, and sport fishing. A particularly interesting feature is the three kilometres of untouched beach and the coral reef that begins just outside the main building.

Palm Tree Village
P.O. Box 327
Scarborough, Tobago
(868) 639-4347/8/9 voice
(868) 639-4180/623-5776 fax

Richmond Great House

This beautifully restored British sugar-plantation mansion is located near the village of Belle Garden on the windward (Atlantic) side of Tobago, high atop a lofty hill that provides fine views in all directions.

This elegant house dates from 1766 and includes five bedrooms, a restaurant, a bar, and a pool. The interior is dark mahogany with high ceilings, huge rooms, and early 20th century furniture. It houses a collection of African sculptures and carvings. The vista is one of flowers, mature fruit trees, rolling coastal hills, and the unbroken forest of the highlands as far as the eye can see.

Richmond Great House
Belle Garden, Tobago
(868) 660-4467 voice

Sea Edge Bed and Breakfast

This private, spacious, ranch-style home is located in Mount Irvine on the leeward (Caribbean) side of Tobago. The view over Buccoo Reef provides magnificent views of brilliant sunsets as well as seasonal sightings of porpoises and marine turtles. The Sea Edge is near Grafton Estate with its abundant birdlife. The Sea Edge is very intimate, offering just two apartments: a one-bedroom and a two-bedroom.

Sea Edge
c/o Scarborough Post Office
Scarborough, Tobago
(868) 639-3926 voice

Speyside Inn

This family-run establishment is located on the windward (Atlantic) side of Tobago on the outskirts of the tiny northern village of Speyside. All five comfortable guest rooms, with terracotta floors, face the Atlantic Ocean. Each room has its own balcony and en-suite bathroom. The Speyside Inn was rebuilt from a beautiful old white-frame house that was owned by the publisher of Trinidad's *Field Naturalist* magazine. Just down the hill is Batteaux Bay across from Little Tobago Island, while behind the inn rise the foothills of Pigeon Peak, Tobago's highest point. The Rainforest Reserve is a mere 20-minute drive away. The owners of the Speyside Inn are conservationists who take pride in their fine cuisine, which is served on the sea grape terrace.

The Speyside Inn
Windward Road
Speyside, Tobago
(868) 660-4852 voice and fax
e-mail: speysideinn@trinidad.net
web-site: http://www.caribinfo.com/speysideinn/

Turpin's (Man o' War Bay) Cottages

Pat Turpin rents six beachside houses near Charlotteville, the northernmost town in Tobago. This is an ideal setting for nature lovers, as the cottages are on the Charlotteville Estate, a 405 ha. (1,000 acre) former cocoa plantation. The entire estate is open to visitors, who may wander in freely from the beach, among cocoa trees and right into the rainforest. From Turpin's Cottages, a five-minute walk along the beach leads to the petite fishing village of Charlotteville, with its shops, markets, rum shops, and restaurants. Turpin's lies between two premier birding attractions, St. Giles (Goat) Island Bird Sanctuary and Little Tobago (Bird of Paradise) Island Bird Sanctuary. The manager arranges and guides birding and nature field trips. Colourful landscaped grounds that comprise a tropical garden of beautiful shrubs and flowers surround the cottages.

The simple cottages, located near the beach, are airy and spacious. There are one-, two-, three- and four-bedroom cottages, each comfortably furnished with a living room, dining area, kitchen with all utensils, a bathroom with hot and cold water, screened and louvered windows, and fans. The cool sea breeze makes air-conditioning unnecessary. In front of each cottage is a veranda with lounges. For a small additional charge a maid can be hired to cook and clean. A dive shop provides snorkelling and scuba equipment. An honour-system commissary provides domestic supplies for those who choose to cook for themselves. Fresh fish is available when the local fishermen bring their catch to the Charlotteville fish house, and local Charlotteville markets supply vegetables and other foodstuffs. Drivers and boatmen are available for excursions throughout Tobago or to nearby St. Giles and Little Tobago Islands.

Turpin's Cottages
Charlotteville, Tobago
(868) 660-4327 voice
(868) 660-4328 fax

OTHER FACILITIES

Visiting birders have reported favourably about the following mostly larger and less nature-oriented properties. For a current list of accommodation, dining places, and other information useful to visitors, request a copy of *Discover Trinidad & Tobago* from Trinidad and Tobago Tourism Services (see page 34).

Port-of-Spain

In Trinidad:

ALICIA'S HOUSE (St. Ann's, in Port-of-Spain)
(868) 623-2802 voice
(868) 623-8560 fax

HALYCONIA INN (Cascade, near Port-of-Spain)
(868) 623-0008 voice
(868) 624-6481 fax

CROWN PLAZA (Port-of-Spain)
(868) 625-3361 voice
(868) 625-4166 fax

KAPOK HOTEL (Port-of-Spain)
(868) 622-6441 voice

LA BELLE MAISON (Maracas)
(868) 663-4413 voice
(868) 642-1076 fax

NORMANDIE HOTEL (St. Ann's, in Port-of-Spain)
(868) 624-1181 voice and fax

PAR-MAY-LA'S INN (Port-of-Spain)
(868) 628-2008 voice
(868) 628-4707 fax

SADILA HOUSE (Arouca)
(868) 640-3659 voice
(868) 640-1376 fax

TROPICAL HOTEL (Maraval, near Port-of-Spain)
(868) 622-5815/4249 voice
(868) 628-3174 fax

VALSAYN VILLA (east of Port-of-Spain)
(868) 645-1193 voice and fax

ZOLLNA HOUSE (Maraval)
(868) 628-3731 voice
(868) 627-0856 fax

In Tobago:

ARTHURS BY THE SEA (near Crown Point)
(868) 639-0196 voice
(868) 639-4122 fax

CHOLSON CHALETS (Mount Pleasant, near Charlotteville)
(868) 639-8553/2847

CONRADO BEACH RESORT (Pigeon Point)
(868) 639-0145/6 voice
(868) 639-0755 fax

GRAFTON BEACH RESORT (Black Rock)
(868) 639-0191 voice
(868) 639-0030 fax

JIMMY'S HOLIDAY RESORT (Crown Point)
(868) 639-8292/8929 voice
(868) 639-3100 fax

PLANTATION BEACH VILLAS LTD. (Black Rock)
(868) 639-0455 voice and fax

SEA BREEZE MOTEL (Scarborough)
(868) 639-6404 voice
(868) 639-5158 fax

BED AND BREAKFASTS

For a current list of bed and breakfasts, contact the following organization:

Tobago Bed and Breakfast Association
c/o Federal Villa
1–3 Rockley Vale
Scarborough, Tobago
(868) 639-3926/8836 voice
(868) 639-3566 fax
Contact: Ann Roberts

CLIMATE AND CLOTHING

The climate is so constant that it is easy to pack the right clothes for a visit. It is recommended that birders wear a wide-brimmed hat for protection against sun and rain. It is particularly important not to venture out in the midday sun without such protection.

Lightweight clothing is appropriate in all seasons, adding a light jacket or sweater for the cool nights, especially at higher altitudes. A long-sleeved shirt is indispensable protection from the sun. Shorts are recommended, although lightweight trousers or jeans are best for exploring savannahs covered with razor grass. A lightweight waterproof poncho or jacket with a hood is useful during the rainy season. A folding umbrella can also be useful especially as in light rain it is possible to continue birdwatching and keep one's binoculars dry. If caught out in a shower the rain will be found to be warm and pleasant unless torrential, and lightweight clothes usually dry rather quickly afterwards. Swimwear is useful as there are many places to swim, and the coral reefs off Tobago are excellent for snorkelling. Swimwear is considered to be inappropriate attire in most hotels and guesthouses.

Trainers (tennis shoes or sneakers) or sports sandals such as Tevas are the footwear of choice most of the time, but do bring an extra pair for use when your wet pair is drying or as foot protection while snorkelling. Hiking boots are very useful on some of the more arduous mountain trails, especially under muddy conditions.

Some mobile (cellular) telephones work in some locations on both islands. Mobile-phone signals are transmitted on a line-of-sight basis so such phones do not work in mountainous areas. Internet cafés can be found on both islands.

Other items to consider bringing include extra batteries, alarm clock, notebook and pencil, daypack, polarizing sunglasses, portable sound recording device (e.g., cassette or minidisc recorder), and plastic zip-locking bags to protect items from dust, moisture, sea air, and ants. Strong sunscreen is essential. A telescope and tripod are very useful, especially when birding along the coast or at wetlands. A camcorder or digital camera is also useful as is a torch and perhaps a spotlight for sighting nocturnal birds and mammals (but be carefully not to shine powerful spotlights directly into the eyes of night creatures). Finally, remember to pack the essentials: binoculars, Richard ffrench's *Guide to the Birds of Trinidad and Tobago*, and this book.

HEALTH & MEDICAL FACILITIES

The drinking water on both islands is safe and need not be boiled or otherwise treated before use.

Chigger

Pest insects such as horseflies, sand flies, and mosquitoes are rarely a problem except at certain periods during the rainy season when insect repellent and anti-itch products are useful. Most biting insects are repelled effectively by any of the commercially available preparations. Chiggers (which are actually mites, not insects) can be common in Trinidad during the dry season, but their bites can be reduced by liberal and frequent application of insect repellent spray to footwear and the lower parts of trousers and legs. To further help avoid them it is wise not to sit on logs in the forest or walk into long grass. Chiggers insert a tube through the skin and inject a proteinaceous anticoagulant that causes a localized allergic reaction that manifests itself several hours later, after the chigger has dropped off. Any anaesthetising sunburn preparation will numb the itching for hours.

There are four species of poisonous snakes in Trinidad – the bushmaster, the fer-de-lance, and two small coral snakes. Snake bites are very rare and fatalities are almost unknown. Most deaths caused by snake bite are of children who have been bitten by one of the coral snakes, the bites of which are not especially painful. The bite goes unreported and untreated and the child dies of respiratory failure a few days later. If a snake is spotted, avoidance is the best course of action. No poisonous snakes occur in Tobago.

Modern health facilities are available on both islands. Mt. Hope Hospital in Trinidad is considered to be among the best in the Caribbean, offering everything from MRIs to recompression chambers for scuba divers. A complete listing of medical facilities in Trinidad and Tobago, with names, addresses, and telephone numbers, can be found at: http://www.tntisland.com/medical.html

The populations of Trinidad and Tobago have increased dramatically in recent years. Tobago is still almost crime-free but the crime rate on Trinidad has jumped alarmingly. Drug-related violent crime is on the increase and usually occurs in heavily populated areas such as Port-of-Spain. Visitors are targeted because they stand out in crowds and often display binoculars, cameras, and jewellery. Most large properties employ security services, at least at night, for the safety of their guests and their buildings and contents. It is essential to be accompanied by an armed guard when venturing with birding groups into one of Trinidad's most productive birding areas, the Port-of-Spain sewage lagoons, because of the proximity of the lagoons to the shantytown along the Beetham Highway. Visiting birders are most often robbed when travelling alone, but even groups should be vigilant. If encountering anyone while on your own it may be worthwhile smiling and saying, 'Respect!' quite loudly as a gesture of goodwill. Entrust all jewellery (including wedding rings), cash, travellers' cheques, and other valuables (such as passports and tickets) to the hotel manager, to be locked in the safe. If leaving valuables in your room cannot be avoided, place them in a locked suitcase or bag. It is wise to keep rooms locked at all times and if anyone knocks, verify their identity before opening the door. When on field trips present a subdued appearance. If it is essential to carry cash, carry it in a hidden pouch or money belt. Do not drape yourself with optical equipment. Try to keep cameras and binoculars out of sight in a bag when not in use. Do not bird alone except on guarded nature reserves such as the Asa Wright Nature Centre or the Pointe-à-Pierre Wildfowl Trust. Wherever possible stay close to your group. Avoid Port-of-Spain and Waller Field at night; the latter is a notorious drug-dealing area. 'Catching a drop' (hitchhiking) is a popular way of travelling in Trinidad and Tobago but it is inadvisable to pick up hitchhikers under any circumstances.

MAPS

The Ministry of Housing & Settlements, Lands and Surveys Division in Port-of-Spain sells detailed road and topographic maps at reasonable prices. The office is located at 18 Frederick Street, Port-of-Spain, near the government's Red House and across the street from the *Trinidad Guardian* newspaper office and NBS radio. Office hours are 8 a.m. to 12 p.m. and 1 p.m. to 2.45 p.m. Monday to Friday. Their telephone numbers are (868) 624-5031, (868) 624-8023, and (868) 627-9201/4.

A comprehensive road map of Trinidad (Edition 4) can be purchased for TT$25.88 from the Lands and Surveys Division, and a road map of Tobago costs about the same. Topographic maps (1:10,000 scale) are available for TT$5. Particularly useful topographic map sections include Sheet 24, covering the Tunapuna-Arima-Piarco quadrant, and Sheet 14, covering the quadrant north of Tunapuna-Arima and including parts of the Northern Range and Blanchisseuse Road. Both maps are in Series E 804 (D.O.S. 316/1). These maps can be obtained in person on weekdays, except on public holidays. Travel bookstores in some of the larger cities in North America and Europe sell maps of Trinidad and a few carry maps of Tobago. Be sure to ask for a road map when renting a car.

Free tourist maps, as well as brochures and advice and assistance in obtaining accommodation, can be obtained at the information offices located at Piarco International Airport (868) 664-5196, Crown Point International Airport (868) 639-0509, and at NIB Mall, Scarborough, Tobago (868) 639-2125/3566.

WHEN TO GO

It is possible for even a novice birder to see between 170 and 225 species of birds in Trinidad and Tobago in two weeks during any month of the year. North American migrants are most abundant from October to March. South American migrants are most abundant from May to September. Consult the bar graph section of this book for information on the seasonal abundance of individual species.

The traditional time for temperate-climate birders to visit Trinidad and Tobago is January to March, coinciding with the driest time of the year. During February the islands pulsate day and night to the sound of steel drums as the nation celebrates the annual holiday of Carnival. At this time many prices are inflated, especially accommodation and transport, including airfares.

Trinidad and Tobago experience a rainy season between late May and late August and another between October and December, but even during the rainy season rain does not fall continuously. September is often a dry month. In the rainy season, torrential downpours occur, with intense, mostly brief, shower activity almost every day, mainly in the early afternoon. Many species forage frenetically between showers and thus are more conspicuous than during the drier times of the year. The rainy season coincides with the Southern Hemisphere's winter and thus is the only time to observe South American species that visit Trinidad and Tobago during the austral winter.

Advantages of visiting during the rainy season include significant reductions in airfares and accommodation, fewer tourists, fewer chiggers, and the presence of South American migrant species. Disadvantages include mosquitoes, the rain, an abundance of mud, and the lack of North American migrants.

June is a particularly good month to visit as sea turtles and seabirds are nesting and South American migrants are present, but it can rain once or twice daily. The first half of May, at the end of the dry season, is a particularly poor time to visit, with parched vegetation and the air hazy with pink dust blown across the Atlantic from the Sahara Desert.

RECOMMENDED BOOKS AND TAPES

Most birders are surprised to learn that James Bond's *Birds of the West Indies* excludes the islands of Trinidad and Tobago from its geographic coverage. Bond and most other ornithologists correctly place Trinidad and Tobago in the Amazonian faunal zone, whose avian distribution differs radically from that of the West Indies.

A Guide to the Birds of Trinidad and Tobago, by Richard ffrench, is an essential item on any birding trip to Trinidad and Tobago. This book provides information on habitat and status, range and subspecies, ringing status, appearance, measurements, voice, food, and behaviour. Be sure to read the introduction, which includes a detailed description of the geography and natural history of the islands. Since there are no illustrations of non-breeding visitors, it would be helpful to bring along one of the many fine field guides to birds of North America to identify wintering or migrating species.

A field guide that covers extralimital species, Steve Hilty's *Birds of Venezuela* is most useful for identifying South American species, especially flycatchers and raptors. It may be used in place of ffrench's guide at a pinch. The greatest drawback to the exclusive use of Hilty's book is the large number of birds illustrated therein that rarely if ever occur in Trinidad and Tobago. Extralimital strays, especially Old World species, are increasingly being reported on both islands, so a field guide to the birds of Europe might also prove useful.

Man is an important part of the natural history of Trinidad and Tobago. For a detailed examination of the human history of Trinidad and Tobago the standard work is Eric Williams's *History of the People of Trinidad and Tobago* (Andre Deutsch, London).

For a current list of relevant literature about Trinidad and Tobago visit the author's website at http//www.inct.net/~billmurphy/books.htm

A cassette recording of 39 species of birds from Trinidad and Tobago can be purchased from the author for US$10 (postage and handling included within the USA). Included are Southern Lapwing, Pale-vented Pigeon, Orange-winged Parrot, Smooth-billed Ani, Ferruginous Pygmy-Owl, Green Hermit, White-tailed and Collared Trogons, Blue-crowned Motmot, Channel-billed Toucan, Pale-breasted and Stripe-breasted Spinetails, Gray-throated Leaftosser, Black-crested and Barred Antshrikes, White-fringed Antwren, Silvered Antbird, Black-faced Antthrush, Tropical Pewee, Great Kiskadee, Fuscous and Boat-billed Flycatchers, Gray Kingbird, Bearded Bellbird, White-bearded and Blue-backed Manakins, Long-billed Gnatwren, Cocoa and White-necked Thrushes, Tropical Mockingbird, Scrub Greenlet, Rufous-browed Peppershrike, Tropical Parula, Bananaquit, Silver-beaked Tanager, Blue-black Grassquit, Yellow-hooded Blackbird, Carib Grackle, and Crested Oropendola.

By studying this tape before visiting Trinidad and Tobago, it is possible to become familiar with many of the frequently heard bird songs of the islands.

ORGANIZED TOURS

The number of people participating in birding trips organised by professional nature tour companies is increasing constantly. The primary advantage of joining such a group is the extreme extent to which pre-tour preparation is reduced, often to the point of simply packing luggage and arranging transport to and from the airport. Another advantage is the opportunity to enjoy the fellowship of kindred spirits. Points to consider include the experience and reputation of the tour company, the expertise of the leaders and local guides, the reputation and convenience of the accommodation, the time of year of the trip, and the cost of the trip in relation to others of a similar nature. Any reputable bird tour company will gladly provide the names of previous tour participants.

Most of the main international bird tour companies offer tours to Trinidad and Tobago in their extensive programmes, and details can be found in their catalogues or on their websites. The author himself leads tours to the islands. Details of these can be found at http://www.inct.net/~billmurphy. In addition the author can vouch for the integrity of the following companies that offer birding trips to Trinidad and Tobago and have particularly close links with the islands:

Caligo Ventures Inc.

Caligo Ventures (named after the caligo owl butterflies, a genus of large eye-spot butterflies found in Trinidad) is the international agent for the Asa Wright Nature Centre. With more than 18 years of experience representing the Centre, Caligo Ventures is uniquely qualified to offer a superior birding experience.

In support of sustainable ecotourism, Caligo uses only local leaders and has partnered with the finest local tour operators to make in-country arrangements, thus ensuring the best possible trips for tour participants.

Included in Caligo Ventures packages are three meals per day; all gratuities except to local leaders, drivers, and boatmen; baggage handling and transportation to and from airports; and a bird checklist. Their tours to Trinidad, based at the Asa Wright Nature Centre, include five field trips throughout the island with some of Trinidad's most experienced local guides. In May and June the field trips also include a nighttime visit to the nesting beaches of the spectacular Leatherback Turtle.

Caligo Ventures' packages also provide ample time to enjoy and explore the excellent birding trails that traverse the Centre's 81-ha (200-acre) sanctuary.

Caligo Ventures' packages that include Tobago offer two additional field trips to the forested home of the elusive White-tailed Sabrewing and Blue-backed Manakin. They also visit Little Tobago Island, one of this nation's most important wildlife sanctuaries, where in the proper season nesting colonies of Red-billed Tropicbird, Red-footed and Brown Boobies, Bridled and Sooty Terns, and Brown Noddies can be seen. Accommodations in Tobago are at the Blue Waters Inn and the Cuffie River Nature Retreat.

For bonafide scientific research projects, facilities and accommodation are available in Trinidad's Arima Valley at the Simla Field Research Station, which is operated by the Asa Wright Nature Centre. Interested parties may submit their requests for information on Simla either to Caligo Ventures or direct to the Centre.

Caligo Ventures Inc.
156 Bedford Road
Armonk, New York 10504, USA
(800) 426-7781 from the U.S. and Canada
(914) 273-6333 from outside the U.S.
e-mail: info@caligo.com
web-site: http://www.caligo.com

Pan Caribe Tours

Paula Beaird is the Director of Pan Caribe Tours, a wholesaler/tour operator that has specialized in tours to Trinidad and Tobago for 19 years with extensive experience in planning tours of all kinds. Pan Caribe Tours puts together trips for travel on any date, with no specific limits on the length of stay on either island. Their tour costs include air, land, and ground transportation. Pan Caribe Tours can customize itineraries to include field trips or any sightseeing that may be required. A potential visitor need only telephone, and the package will be made up at the lowest overall cost. For those who prefer to work through their favorite travel agent, Pan Caribe will work directly with your agent for the same low rate. Office hours are 8.30 a.m. through 5.30 p.m. Central Standard Time and 10 a.m. through 1 p.m. on most Saturdays.

Pan Caribe Tours
P.O. Box 3223
Austin, Texas 78764 USA
(800) 525-6896 (USA and Canada) voice
(512) 266-7995 (elsewhere) voice
(512) 266-7986 fax
e-mail: info@pancaribetours.com
web-site: http://www.pancaribetours.com/

Tropical Birding Adventures Inc.

Another nature tour operator that organizes birdwatching trips to Trinidad and Tobago is Tropical Birding Adventures, established by an avid and enthusiastic birder, Nancy Akosa. Nancy wholeheartedly believes in the importance and value of working with local expert birding guides such as Courtenay Rooks. As one might expect, native

Trinidad and Tobago guides such as Courtenay are the ones most familiar with the presence and location of any unusual or exotic species. Tropical Birding Adventures specializes in two types of tours: personalized itineraries for independent birders who prefer not to travel with groups, and group tours. Participants have all preparations and details handled for them from arrival through departure, with choices of accommodation, meal plans, and length of tour.

One special field trip is offered on both tours: a chance to see the rare Trinidad Piping-Guan, a species very rarely seen by birders visiting Trinidad.

Tropical Birding Adventures offers its clients the option to split their Trinidad accommodation between the Asa Wright Nature Centre and alternative accommodation south of the Northern Range. When visiting sites west of the Arima Valley the use of alternate accommodation maximizes birding time by eliminating the 45-minute drive to and from the Asa Wright Nature Centre.

Tropical Birding Adventures
4534 North Kasson
Chicago, IL 60630, USA
Toll free: (800) 462-2473
(773) 777-2923 voice and fax
e-mail: scarletibis2000@earthlink.net

PHOTOGRAPHY

Most people bring some kind of camera with them on their visits to Trinidad and Tobago. Besides capturing memories, recording equipment is indispensable for documenting rarities. Using a video or digital camera is straightforward; just be sure to keep batteries charged. The use of optical cameras requires much more forethought. For photographing scenery, any lens from 28 mm to 200 mm is fine. For photographing birds, a minimum focal length of 300 mm is needed and 500 mm is even better. Consider bringing a medium-length lens with a 2X or 3X multiplier. For those who like to photograph at dawn or dusk, it might be helpful to know that the day length in Trinidad and Tobago at any time of year averages 12 hours, with sunrise and sunset being within about an hour on either side of 6 a.m. and 6 p.m., respectively. Dawn and dusk are of surprisingly brief duration throughout the tropics.

In general, techniques applicable in temperate zones work in Trinidad and Tobago except that a daylight filter must be used to counteract the brilliant blue component in the sunlight. Even though tropical sunlight is intense, and slow film (ASA 64) is perfectly fine for most outdoor shots, at noon an *f*16 day in full sunlight may be an *f*1.8 day in the rainforest. Most photographers therefore carry film ranging in speed from ASA 64 to ASA 1000. Photographic film is available in every major town. High-quality photographic processing is available in Port-of-Spain and in other large cities.

Photographing Scarlet Ibis flying in to their roost at dusk is a serious challenge, no matter what kind of camera used. For this endeavour, using a digital or video camera is recommended; if using an optical camera, use the fastest film possible. When photographing the Scarlet Ibis, three factors have to be contend with – a gently rocking boat, a quickly diminishing light level, and a moving target. Few visitors achieve success in this arena but you may be one of the lucky ones. Please do not ask the boatman to violate the minimum-approach markings near the roost sites or to race the outboard motor in an effort to agitate the birds to flight. Such action is detrimental to the ibis and to birders coming on future visits.

ATTRACTING BIRDS TO THE OBSERVER

Obvious though it may seem, the more an observer blends in with the environment, the closer the birds will come and the more birds will be seen. Dress to be inconspicuous, preferably in greens or browns. Remember that you are an intruder in an area the birds know well. The less movement or sound, the better the chances of observing secretive species.

While 'pishing' may attract birds in the Northern Hemisphere, it often does not work in Trinidad and Tobago. Although some birds may occasionally respond to such a sound, far more will flock to the call of the Ferruginous Pygmy-Owl. To imitate this species purse your lips and whistle 'hoo' about four times per second. The pitch does not matter as much as the tempo. The genuine call will be heard soon after reaching Trinidad and imitations can then be perfected that will often attract the owl itself. Recordings of other species, made on the spot and played back, are also very effective at bringing skulking species into view, but recordings should be played only until the bird appears. Do not harass a bird by lengthy playbacks of its call – the bird may become very stressed and may even abandon its territory. Playback is prohibited in the grounds of the Asa Wright Nature Centre due to the constant presence of so many birdwatchers.

From December to March it can be worth imitating the call of the Eastern Screech-Owl, which attracts North American migrants. Wait until the birds have tired of the Ferruginous Pygmy-Owl imitation and then try the screech-owl call. Often two or three species such as American Redstart, Prothonotary Warbler, and Northern Waterthrush will hop out of the bushes and into view after having ignored your earlier efforts.

One little-known fact is that many tropical species of birds are quite sedentary, using a 'sit-and-wait' hunting style. When the preferred insect or other prey appears they dive quickly, snatch the prey, and return to the same perch or to another perch nearby. If the few seconds of movement is missed then that species of bird may be missed too. As an example of the kind of payoff that patient watching can yield, consider that virtually every White-throated Spadebill the author has ever seen has been spotted through the 'sit-and-wait' method, remaining still and watching for a sudden movement. It is very useful to acquire at least this one new birding technique to develop good tropical birding skills.

TRINIDAD AND TOBAGO TOURIST BOARD OFFICES

Additional information about Trinidad and Tobago may be obtained from the following offices:

Head Office in Trinidad

Trinidad and Tobago Tourism Development Authority
134-138 Frederick Street
Port-of-Spain, Trinidad
(868) 623-1932/4; 623-INFO voice
(868) 623-3848 fax

United Kingdom

(Office closed; for general information contact the British Embassy, listed below)

United States

Trinidad and Tobago Tourism Services
c/o Sales, Marketing and Reservations Tourism Services (SMARTS)
7000 Boulevard East
East Guttenberg, New Jersey 07093, USA
(201) 662-3403 voice
(201) 869-7628 fax

Trinidad and Tobago Tourism Services
c/o Cheryl Andrews Marketing Inc.
1500 San Remo, Suite 145
Coral Gables, Florida 33146, USA
(305) 663-1660 or (800) 595-1TNT voice
(305) 666-9723 fax

Canada

Trinidad and Tobago Tourism & Trade Center
York Center
145 King Street West & University Avenue
Toronto, Ontario M5H 1J8, Canada
(416) 367-0390

EMBASSIES OF THE UNITED KINGDOM, THE UNITED STATES, AND CANADA IN TRINIDAD AND TOBAGO

United Kingdom

Furness House
90 Independence Square
Port-of-Spain, Trinidad
(868) 625-2861

United States

15 Queens Park West
Port-of-Spain, Trinidad
(868) 622-6371

Canada

Huggins Building
72-74 South Quay
Port-of-Spain, Trinidad
(868) 623-4787/7250

TRINIDAD AND TOBAGO EMBASSIES AND HIGH COMMISSIONS IN THE UNITED KINGDOM, THE UNITED STATES, AND CANADA

United Kingdom

High Commission for Trinidad and Tobago
42 Belgrave Square
London SW1X 8NT, United Kingdom
020-7245-9351 (voice)
020-7823-1065 (fax)
e-mail: tthc@virgin.net

United States

Embassy of Trinidad and Tobago
1708 Massachusetts Avenue, NW
Washington, DC 20036, USA
(202) 467-6490

Permanent Trinidad and Tobago Mission to the United Nations
801 Second Avenue
New York, New York 10017, USA
(212) 697-7620

Canada

Consulate for Trinidad and Tobago
P.O. Box 1526, Station B
Ottawa, Ontario, K1P 5R5, Canada
(613) 232-2418

High Commission for Trinidad and Tobago
365 Bloor Street East, Suite 1700
Toronto, Ontario, M4W 3L4, Canada
(416) 922-3175

INTRODUCTION TO SITE INFORMATION

From an avifaunal viewpoint Trinidad is easier to divide into birding regions than is tiny Tobago. Thus for Trinidad the sites herein are presented in terms of topography – the Northern Range, the foothills and savannahs, and the wetlands. Each region will present the birder with a large number of species found only in that habitat as well as with some 'overlap' species from other habitats.

Tobago is best presented from a straight geographic viewpoint beginning at Crown Point International Airport in the extreme southwest and ending at Little Tobago Island in the extreme northeast – Crown Point International Airport Area, Pigeon Point, Bon Accord Treatment Ponds, and Buccoo Marsh in the southwest; Grafton Estate and Arnos Vale on the west coast; Belmont Road, Hillsborough Dam, and Old Castara Road on the east coast; and Louis D'Or River and Beach, and Little Tobago ('Bird of Paradise') Island in the northeast.

As mentioned earlier, bird names and their arrangement generally follow the American Ornithologists' Union's Check-list of North American Birds, 7th edition, 1998 (Am. Ornithol. Union, Washington, D.C.).

Access to many of the sites described herein changes depending on factors such as road construction, changes in land ownership and use, and the current government's security policy. Before driving to a birding site it is highly recommended that you discuss the condition of the current site and the directions to it with the hotel manager or with any of the local contacts listed on page 156.

BIRDING IN TRINIDAD

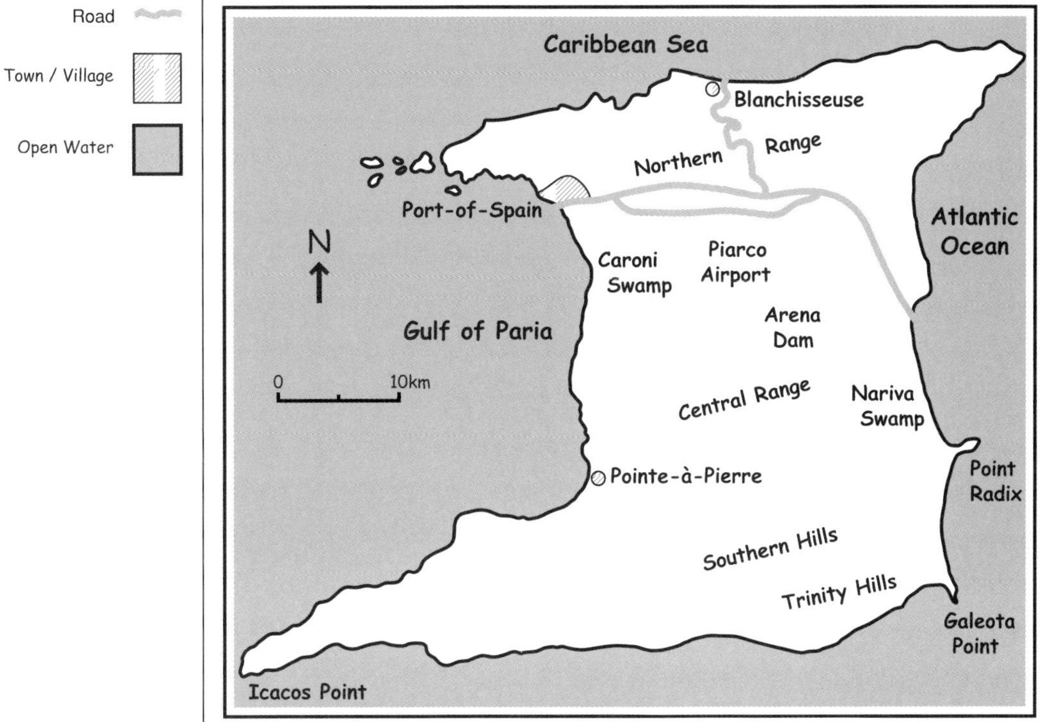

Provided here are detailed descriptions of birding trips to different parts of Trinidad. For a visit of a week's duration, a suitable itinerary might include the following:

Day 1 Grounds of hotel or guest house and environs – learn most of the local birds
Day 2 Blanchisseuse Road from Arima at least as far as Brasso Seco Trace
Day 3 Asa Wright Nature Centre/Oilbird cave, Lalaja Trace
Day 4 Waller Field, Aripo Savannah
Day 5 Arena Forest and Nariva Swamp
Day 6 Trincity Ponds, Caroni rice fields, Caroni Swamp

In the site descriptions the numbers following the locations are approximate distances driven during the course of that foray and the total anticipated amount of travel and birding time. Most distances and heights are given in kilometres and metres.

As many independent birders visiting Trinidad stay at accommodation east of the capital city, the starting points given are east of Port-of-Spain.

Ubiquitous Species

The following species are so common, widespread, and conspicuous that little further mention is made of them: Cattle Egret, Black and Turkey Vultures, Ruddy Ground-Dove, Orange-winged Parrot, Smooth-billed Ani, Short-tailed Swift, Yellow-bellied Elaenia, Tropical

Pewee, Great Kiskadee, Boat-billed Flycatcher, Tropical Kingbird, Bare-eyed Thrush, Tropical Mockingbird, Bananaquit, Blue-gray and Palm Tanagers, Carib Grackle, Blue-black Grassquit, and Yellow Oriole.

The phrase 'the usual species' refers to those species that respond readily to the call of the Ferruginous Pygmy-Owl: Rufous-breasted and 'Tropical' House-Wren, Bare-eyed Thrush, Rufous-browed Peppershrike, Red-eyed ('Chivi') Vireo, Golden-fronted Greenlet, Bananaquit, Blue Dacnis, Green, Purple, and Red-legged Honeycreepers, Turquoise, Bay-headed, Blue-gray, Palm, Silver-beaked, White-shouldered, and White-lined Tanagers, and Violaceous Euphonia.

NORTHERN RANGE

Asa Wright Nature Centre
30 km – 1 day

To reach the Asa Wright Nature Centre follow the Eastern Main Road to the town of Arima. On entering Arima from the west cross a concrete bridge and pass a sports arena (velodrome) on the right. At the first set of traffic lights there is a petrol station on the left, a bus depot on the right, and road signs in front that read 'No Entry Ahead'. At the second set of traffic lights bear left, passing another petrol station on the right. After three blocks turn right at a stop sign, onto Broadway (no street sign). Two blocks farther on in the town centre there is a clock on a pole in the middle of the road. Pass to the left and take the first left, St. Joseph Street (no street sign). Follow this road for 1.4 km to its junction with the Arima Bypass (North Demerara Road), which merges from the right. The city of Arima ends abruptly here and the rainforest begins. Follow Blanchisseuse ('washerwoman') Road into the Northern Range.

An alternative route, good during heavy traffic, is to follow the Churchill-Roosevelt Highway east, turning left at the Arima exit at Demerara Road. Stay on Demerara Road north to the Eastern Main Road, then turn right onto the well-marked Blanchisseusse Road. Turn left, thus avoiding most of the congestion in downtown Arima. The environment changes on leaving the city of Arima. Continue for about 11 km and turn left at the well-marked entrance to the Asa Wright Nature Centre. Follow the rustic asphalt road to the visitors' car park. The estate house is down thc hill.

The Asa Wright Nature Centre is located on a ridge side in the Arima Valley at an elevation of 360 m (1,181 ft.), 12 km (7.5 miles) north of the city of Arima (an Amerindian name meaning 'place of water'). A former coffee-cocoa-citrus plantation known as Springhill Estate, the Asa Wright Nature Centre has been allowed to return to a wild state and is a grand place to spend some time. Specialities of the Asa Wright Nature Centre include:

Switchback on Blanchisseuse Road

White Hawk, Ornate Hawk-Eagle, Blue-headed Parrot, Spectacled Owl, Oilbird, Tufted Coquette and about 10 other species of hummingbirds, all three species of trogons, Golden-olive and Chestnut Woodpeckers, Black-tailed Tityra, Bearded Bellbird, a typical ground-level lek of White-bearded Manakins, most of the tanagers and their allies, and a nesting colony of Crested Oropendolas.

A daily fee of US$6 is payable at the front desk, and there is a visitors' book to sign. The Centre offers a wide selection of field guides, trail guides, and other books, gifts and local crafts. Guided walks to Dunstan (Oilbird) Cave and other places can be arranged here. Note that one must reside as a guest at the AWNC for a minimum of three nights to be granted access to the Oilbird cave. There is a logbook near the veranda giving details of recent sightings. The Centre also provides lunch, which costs US$6.50 at the time of writing.

From the veranda the entire length of the Arima Valley can be surveyed and flying Channel-billed Toucans, White Hawks, and Ornate Hawk-Eagles can be studied. It is also possible to obtain very close views and photographs of hummingbirds, tanagers, and honeycreepers at the fruit-filled feeders. Nature trails radiate from the estate house towards several different habitats – along and into the lush rainforest, along the bird-rich entrance road, and to Dunstan Cave, home of the famous Oilbird. One very easy and rewarding stroll is along the 1.5-km (roundtrip) entrance road. Blue-headed and Orange-winged Parrots, Squirrel Cuckoo, Great Antshrike, Cocoa, Bare-eyed, and White-necked Thrushes, and Crested Oropendola can all be seen here. Often a Tegu lizard forages beneath the fruit-filled bird feeders.

Tegu lizard

One of the most popular trails leads downhill from the veranda and passes through a citrus grove where trogons and thrushes can be seen. Toucans feed in a towering nutmeg tree while Bearded Bellbirds call 'BOK!' from the rainforest on the right. It is necessary to be accompanied by an authorized guide as access to this trail is restricted. At the far side of a grassy clearing, just past a gigantic mango tree, a clearly defined path marked 'No Entry' leads off to the right. This fairly steep but fascinating trail leads about 2 km through a rainforest frequented by species with such unusual names as Streaked Xenops and Gray-throated Leaftosser. The trail ends at the Guacharo River, at the mouth of Dunstan Cave. From the mouth of the cave birders usually can see a few Oilbirds, but do not disturb them by entering the cave without a guide.

While in the ravine, watch for scintillating blue morpho butterflies and inspect the cliff walls for nesting Chestnut-collared Swifts.

Another short but productive trail leads past the bungalows above the estate house and connects with the entrance road (0.5 km). The forest on the left is home to Black-faced Antthrush, whose imitated three- to five-note descending whistle will often bring these rail-like birds into sight.

Blanchisseuse Road
52 km – 1 day

Arima Valley from Lalaja

This trip starts in the town of Arima and travels north along Blanchisseuse Road to the village of Blanchisseuse on the Caribbean coast, or at least as far as Brasso Seco Trace if a shorter trip is desired. The route winds through some of the most picturesque mountains and highest rainforest on the island. Drivers should take care as the road is hazardous – there are 387 bends and switchbacks with only wrist-thick bamboo guard-rails separating the road from the almost vertical drops. The views are breathtaking, with waterfalls and lush vegetation that includes elephant ear and philodendron as well as ferns and mosses.

To reach Blanchisseuse Road follow the directions to the Asa Wright Nature Centre as far as downtown Arima. From the intersection of St. Joseph Road and the Arima Bypass continue straight ahead into the mountains. The Arima River on the right (0.5 km) is a good place to look for Gray-headed Kite and kingfishers. Pass the Hindu settlement of Verdant Vale (4.3 km) and pull off on the left 0.3 km farther on. A British-colonial-style concrete bridge marks the beginning of an area especially productive for rainforest birds such as Little Tinamou, Lilac-tailed Parrotlet, Blue-headed Parrot, all three species of hermits, Copper-rumped Hummingbird, all three species of trogons, Blue-crowned Motmot, Ochre-lored Flatbill, Black-tailed Tityra, both species of manakin, Rufous-breasted Wren, Long-billed Gnatwren, and Crested Oropendola. Walk a short distance off the road into the forest on the left and imitate the call of the Ferruginous Pygmy-Owl to attract the usual species. It is also possible to venture down the grassy lane on the right side of the road on foot for about 0.5 km.

Between December and March the mountain immortelle trees are in fiery bloom, colouring the rainforest canopy bright orange and attracting a great variety of nectar-feeding birds and butterflies. It is possible to observe all three species of trogons along this route as well as White and Gray Hawks, Channel-billed Toucan, and Golden-headed and White-bearded Manakins. The last two species often display on their communal courts ('leks'), the White-bearded Manakins at ground level and the Golden-headed Manakins 10 m (40 ft.) above ground level. It is also possible to see Black-faced Antthrush, Bearded Bellbird, Golden-crowned Warbler, Speckled Tanager, and Swallow-Tanager in the heights. Other species that are often easier to see along Blanchisseuse Road than anywhere else in Trinidad include Ornate Hawk-Eagle, Green Kingfisher, and Gray Kingbird. The last two species are mainly seen in the lowlands along the Caribbean coast.

Crested Oropendola nests

Drive on and turn right at a sign that says 'Ramharry Garibdass Bros. Contracting Ltd.' (0.3 km). This rough track serves as the entrance road to the Simla Research Station. Founded by the famous naturalist William Beebe, Simla is now managed by the Board of Directors of the

Asa Wright Nature Centre. Turn left at the gated barrier at a sand quarry (0.4 km), park, and explore the rest of the entrance road on foot.

Then drive to the Simla carport (0.6 km) and walk up the stairs on the left to the main building. Shout greetings to make your presence known to the other guests. Trinidad Euphonia can often be found more easily at Simla than elsewhere. They frequently feed in clumps of mistletoe. Double-toothed Kites have nested in the trees above the carport. Explore the entire Simla area and the trails leading from the main buildings into the rainforest.

On leaving Simla return to Blanchisseuse Road and turn right. Stop at as many of the numerous pull-offs as time permits. Turn right on reaching Lalaja South Road (3.0 km). This road is steep in places but the views of the Arima Valley alone are worth making this trip. The many overlooks are great for spotting raptors such as Swallow-tailed Kite and Bat Falcon. Walk along Cooker Trace, which leads off on the right for 0.3 km to its end at a house (0.6 km), searching in particular for Squirrel and Striped Cuckoos. Swifts often cross the ridge top here at knee level, moving from the Arima Valley to the Guanapo Valley and back. Most of them are Band-rumped or Gray-rumped, but there may be other species mixed in with them.

Ahead 0.2 km is a spectacular view of the sparsely inhabited Guanapo Valley to the east; this overlook is particularly good for raptors, with Gray-headed and Swallow-tailed Kites, White Hawk, Common Black-Hawk (often in pairs), and Bat Falcon among the more common species. Scaled Pigeons and White-tipped Doves often streak across the valley and Channel-billed Toucans are unmistakable with their woodpecker-like glides.

Lalaja Bridge Trace ahead 0.5 km is too steep for vehicles except those with four-wheel drive but it provides an invigorating walk to a forest clearing.

Across a rustic bridge (1.3 km), Vincent Trace leads off on the right (0.4 km), another good birding area. Ahead 0.3 km, Lalaja-Paria Trace leads off to the left. Farther ahead, Lalaja Road deteriorates into an unpaved, rutted path that winds through semi-abandoned coffee-cocoa estates that abound with nesting colonies of Crested Oropendola. Turn around at Lalaja-Paria Trace and return to Blanchisseuse Road.

Proceed 2.3 km farther up Blanchisseuse Road and scan the rainforest halfway up the far side of the valley for the Asa Wright Nature Centre. The Asa Wright Nature Centre can be reached via an entrance road on the left (0.3 km).

Pendulous nests of Crested Oropendolas overhang the road 1.1 km ahead. Stop at the power-line clearcut 1.7 km farther and enjoy a splendid view of the Arima Valley and the Arena Reservoir far out in the distance and to the right. This overlook is a good spot for observing White Hawk, both species of parrots, and Channel-billed Toucan. Park at the entrance to the Textel microwave station on the right (0.4 km).

Search the low brush along the entrance drive for Speckled Tanager, which is frequently seen at this spot. Drive to the top and ask permission to bird the summit, over which Lesser Swallow-tailed Swifts and raptors can be seen. Back on Blanchisseuse Road, about 0.5 km farther is a good vantage point on the left for viewing the Northern Range Forest Reserve 'A' and the Lopinot Valley to the west. Slaty-capped Flycatcher is often seen here.

Las Lapas Trace and Andrews Trace are on the left across the road from a fenced-in transmitting station (0.9 km). The elevation at this point is 595 m, the highest elevation on Blanchisseuse Road. Andrews Trace runs essentially east-west for several kilometres in either direction maintaining an elevation of 455-760 m. Plan to spend an hour here looking for the high-elevation specialities – Orange-billed Nightingale-Thrush, Yellow-legged Thrush, and Blue-capped and Speckled Tanagers – as well as looking and listening for Tropical Parula and Golden-crowned Warblers, both of which are common. Birders have seen or heard Spectacled Owl here in recent years. From October to March look for migrant North American warblers such as American Redstart and Northern Waterthrush.

A visit to Paria Springs, a secluded estate 305 m above sea level, is recommended. Continue along Blanchisseuse Road, which at this point heads down the windward side of the Northern Range. At a switchback ahead (1.4 km) is a sign that says 'Paria-Morne Blue [sic] Road, Brasso Seco 4 km, Blanchisseuse 12 km'. To reach Blanchisseuse it is necessary to turn hard left here, but to get to Paria Springs continue on the Paria/Morne Bleu Road for about 9.6 km into the village of Brasso Seco/Paria. Red Howler Monkeys are rare here but sometimes can be heard along this stretch of the road. Turn right at the Information Centre onto Madamas Road. Drive through the village, past the rum shop and church on the left. Take the next right, onto the rough but driveable Gomez Trace, which is 1.6 km long and ends at Paria Springs Estate. (At time of writing a large landslide had closed Gomez Trace 1.2 km from its intersection with Madamas Road but visitors could still walk Gomez Trace from that point.) Birding is excellent all along the road. At the estate look for a small house and a wild cherry tree about 6 m (20 feet) before the end of the road. The cherry tree is almost always filled with birds, especially Bay-headed Tanagers. Trails on the estate provide further birding opportunities.

Red Howler Monkeys

Among the 140+ species recorded at Paria Springs, Blue-headed Parrots are common with as many as 40 being found in a single fruiting tree.

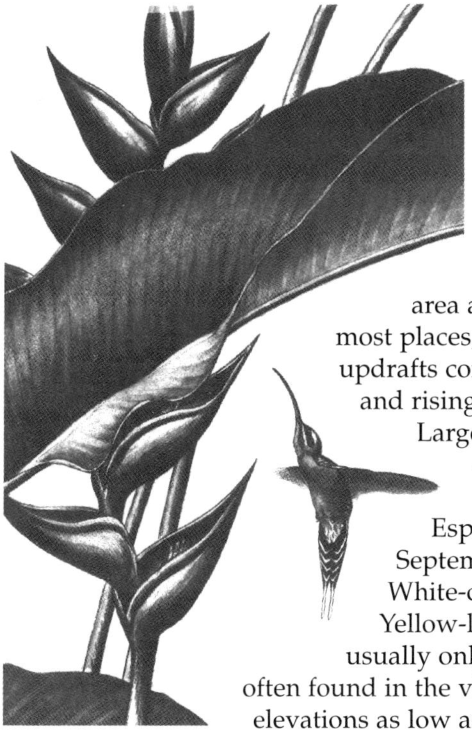

Green Hermit

The area between Morne Bleu and Paria Springs seems to attract more warblers than most other areas. Golden-rumped Euphonia is not easy to find anywhere in Trinidad but it shows up now and again around Paria Springs. The area also attracts more raptors than most places, possibly a result of the steady updrafts coming off the windward shore and rising over El Cerro del Aripo. Large numbers of Swallow-tailed and Plumbeous Kites can be seen riding these updrafts. Especially in August and September, flocks of hundreds of White-collared Swifts spiral upwards. Yellow-legged Thrush, a species usually only found at high elevations, is often found in the vicinity of Paria Springs at elevations as low as 150 m.

Other birds of special interest found at Paria Springs include White and Short-tailed Hawks, Ornate Hawk-Eagle (occasional), Lesser Swallow-tailed Swift, Blue-crowned Motmot, Channel-billed Toucan, Crimson-crested Woodpecker, Black-tailed Tityra, Bearded Bellbird, Yellow-legged Thrush, Yellow-rumped Cacique, and Golden-crowned Warbler. Flycatchers in the area are surprisingly varied, including uncommon species such as Small-billed Elaenia, Yellow-Olive and Olive-striped Flycatchers, and White-winged Becard.

For those wishing to stay in Paria Springs, accommodation for birders is available along the Paria/Morne Bleu Road, in the village of Brasso Seco/Paria Springs, and on Gomez Trace. For details, contact Courtenay Rooks (see page 159)

Return to the junction of Paria/Morne Bleu and Blanchisseuse Roads and bear right. The route is all downhill for the next few kilometres. The habitat becomes drier and more typically Caribbean as you proceed. Continue 1.6 km to a seasonally open grassy area good at times for the nearly extirpated Yellow-bellied Seedeater, the uncommon Sooty Grassquit, and recently for a lone Indigo Bunting. Not far ahead (1.1 km), just beyond the spot where Morne La Croix Trace leads off to the left, there is a distant view of the hamlet of Morne La Croix itself far below. El Brasso Trace leads off on the right (0.6 km); this is a good trail to walk for ant-associating birds such as Plain-brown and Cocoa Woodcreepers, Great Antshrike, Plain Antvireo, and Black-faced Antthrush.

The junction where Morne La Croix Trace rejoins Blanchisseuse Road on the left can be good for Band-rumped Swift (1.2 km). Within sight on the left ahead (0.1 km), in the small hamlet of Morne La Croix, is the famous hedge of blue vervain that the travel brochures mention as frequented by Tufted Coquette, a species unlikely to be seen here. Ahead 2.5 km on the left is another wonderful overlook to the west, good for White, Common Black-, and Zone-tailed Hawks and for viewing swifts from above, the best angle from which to identify them. Collared Trogon is often seen 4.1 km farther ahead in a delightfully cool, shady valley at a bend in the Marianne River.

Blanchisseuse Beach

Old Blanchisseuse Road leads off on the left in 5.0 km. Keep to the right as the road climbs the Andes Mountains (5.4 km). The Caribbean lies directly ahead. On reaching the North Coast Road (1.6 km), turn right and then bear left at 0.9 km at the Y-junction. A bathing beach on the left offers a panoramic view of the Caribbean, where Brown Pelicans and Magnificent Frigatebirds are abundant (0.7 km). Farther down the road a cable bridge crosses the mouth of the Marianne River. A sandy sheltered lagoon provides habitat for Solitary and Spotted Sandpipers as well as for Green and American Pygmy Kingfishers (0.4 km). Among the species flycatching over the bridge are Short-tailed Swift, Gray-breasted Martin, and Southern Rough-winged Swallow. A trail on the right just before the bridge leads into low-lying woodlands, good for Tropical Screech-Owl and Ferruginous Pygmy-Owl. King Vultures, both adult and immature, have been seen soaring high over this region in recent years.

Retrace your route 2.0 km to the intersection of Blanchisseuse and North Coast Roads. Just ahead Lionel Oliver's store on the left offers drinks and especially good currant rolls (0.1 km). Scan the power-lines in Blanchisseuse for Gray Kingbird, a species more common here than anywhere else in Trinidad. A right turn 2.2 km farther, onto an inconspicuous dirt lane just after a small bridge, provides access to a tiny photogenic beach nestled between rocky headlands. The Damier

River at the east (right) end of the beach is a reliable site for finding American Pygmy Kingfisher.

Damier River

After leaving the beach turn left onto Blanchisseuse Road and drive about 0.5 km east. Old Blanchisseuse Road is on the right, just before the police station. This excursion provides an exhilaratingly steep climb and superb views of the Caribbean from the top. It is possible to spot the island of Tobago far off in the distance and on the right about 50 km (31 miles) away.

Heights of Aripo
32 km – 1 day

It is recommended that this trip be started early enough to reach the foothills of the Northern Range at first light to experience the rainforest as it awakens and when the maximum number of species are active.

To reach the Heights of Aripo area drive east from Arima on the Eastern Main Road passing the Aripo Livestock Station on the left. Proceed for 1.8 km to the Heights of Aripo Road. A sign on the left on the far side of the junction says 'To ARIPO VILLAGE 11.2 km'. Turn left and begin birding immediately, pulling off the narrow road wherever possible, looking and listening for the usual rainforest species as well as for Little Tinamou and Black-faced Antthrush.

El Cerro del Aripo is the highest mountain peak in Trinidad, reaching an elevation of just over 915 m (3,010 ft). Species of birds generally found only above 600 m in Trinidad include Orange-billed Nightingale-Thrush and Blue-capped Tanager. Other species of special interest include White and Zone-tailed Hawks, Bat Falcon, the rare Trinidad Piping-Guan, Chestnut-collared, Band-rumped, Lesser

Swallow-tailed, and Gray-rumped Swifts, all three species of hermits and the rare Brown Violet-ear, Collared and White-tailed Trogons, Chestnut and Lineated Woodpeckers, Gray-throated Leaftosser, Plain Antvireo, Black-faced Antthrush, Short-tailed Pygmy-Tyrant, Bright-rumped Attila, Black-tailed Tityra, Bearded Bellbird, Yellow-legged, Cocoa, and White-necked Thrushes, Red-crowned Ant-Tanager, and Swallow-Tanager.

Lesser Swallow-tailed Swifts

Follow the Heights of Aripo Road uphill, ignoring all lesser roads leading away from this paved but very winding road. At 6.5 km you will reach a bridge on the right that crosses the Aripo River. Continue straight ahead onto L'Orange Road and enter a former coffee-cocoa-citrus plantation. Growing along the road are breadfruit and breadnut trees. Turn right on Ruiz Trace (0.4 km), a startlingly steep but paved road that leads up to a prime Bearded Bellbird display area. Drive as far as you dare but park before the tarmac ends. The road is so steep that brakes will not hold cars on the grass near the summit.

Bird on foot searching especially for Bat Falcon, Trinidad Piping-Guan, Blue-headed and Orange-winged Parrots, Channel-billed Toucan, Bright-rumped Attila, and Black-tailed Tityra. In the proper season all of the species mentioned at the beginning of this section can be found here.

Return to L'Orange Road, turn left, and continue for 0.7 km. Park just before crossing a small green-and-white concrete bridge. Walk and bird the stretch of road ahead for about 1.6 km, looking for all three species of trogons, Green and American Pygmy Kingfishers, Channel-billed Toucan, thrushes, tanagers, euphonias, and the other usual rainforest species. The estate house on the left just past the bridge attracts at least 11 species of hummingbirds: Little, Green, and Rufous-breasted Hermits, the rare Brown Violet-ear, White-necked Jacobin, Black-throated Mango, Ruby-topaz, Blue-chinned Sapphire, White-chested Emerald, Copper-rumped, and Long-billed Starthroat. The larger trees on the right sometimes hold roosting Tropical Screech-Owl and Ferruginous Pygmy-Owl.

Ruiz Trace

Retrace your route to the junction of L'Orange and Heights of Aripo Roads at the bridge over the Aripo River. Turn left and cross the bridge onto the Heights of Aripo Road. Check the watercress ponds on the right for Green and American Pygmy Kingfishers and for the occasional migrant shorebird. Pass through the village of Aripo. At 4.1 km park at the obvious trailhead on the left (Dandrade Trace). Ignore the private road on the right, which continues only a little farther uphill. Dandrade Trace leads 15 km uphill and down through magnificent rainforest. Because the birding here can be so productive, plan to spend as much time walking as possible. Many of the ant-loving bird species can be found here as well as less specialized migrant species.

If you plan to visit the Caves of Aripo to see Oilbirds, hiring a local guide in Aripo is recommended – signs along the Heights of Aripo Road advertise several. Alternatively, use the site guide in one of the *Nature Trails of Trinidad* books (obtainable at the Asa Wright Nature Centre), but note that trails become obscured quickly in the tropics and may be obliterated by fallen branches and other debris. A local guide can be a wise investment in such circumstances.

Hollis Reservoir
32 km – 1 day

As with the Heights of Aripo foray, this trip is recommended for the early morning when the maximum number of species are active.

A permit is required to enter Hollis Reservoir. It must be obtained in advance from the Water and Sewerage Authority (WASA), Farm Road, St. Joseph, Trinidad, for TT$5.50.

Hollis Reservoir is the largest body of water in the Northern Range. It lies about 4 km north of the town of Valencia. Its entire catchment is undisturbed so the pristine rainforest surrounding it hosts many species. Birds to look for include Little Tinamou, Least Grebe, Anhinga, White Hawk, Broad-winged Hawk, Ornate Hawk-Eagle (nests on the eastern side), Scaled Pigeon, Green-rumped and Lilac-tailed Parrotlets, Blue-headed and Orange-winged Parrots, Squirrel Cuckoo, Rufous-breasted and Little Hermits, Black-throated Mango, all three trogons, Blue-crowned Motmot, Belted and Green Kingfishers, Channel-billed Toucan, Chestnut and Lineated Woodpeckers, Gray-throated Leaftosser, White-flanked Antwren, White-bellied Antbird, Black-tailed Tityra, Bearded Bellbird, White-bearded and Golden-headed Manakins, White-winged Swallow, White-necked Thrush, Rufous-browed Peppershrike, Blue Dacnis, Red-crowned Ant-Tanager, and Yellow-rumped Cacique.

To reach the reservoir travel east from Arima along the Eastern Main Road to the town of Valencia. Follow Valencia Road to the left, leaving the Eastern Main Road, which continues to the right. Travel 0.5 km, ignoring two smaller roads on the left. On the right 100 m beyond a small church is a road on the left that leads to Hollis Reservoir. Travel north along this road, ignoring all side roads. After 1.1 km there is a checkpoint where permits must be shown.

Proceed for 4.0 km along a winding road that runs alongside the Quare River. On passing through a gate there are a series of small buildings on the left. The last small building is the checkpoint where permits must be shown.

Proceed another 0.8 km to the southern end of the Hollis Reservoir. Park in front of the metal railing outside a small storage shed (with toilets) and the staff house. Walk on the top of the dam and scan the area for Least Grebe, Anhinga, White Hawk, Ornate Hawk-Eagle, Scaled Pigeon, parrots and parrotlets, trogons, Belted Kingfisher, Channel-billed Toucan, White-winged Swallow, and Yellow-rumped Cacique.

Continue along the narrow footpath, keeping the reservoir on the right, past the storage shed. Start birding immediately in the forest edges next to the path and also scan across the reservoir (this is very effective because the path is elevated to the equivalent of treetop level on the far side). Watch for Little Hermit, Blue-chinned Sapphire, Golden-headed Manakin, Rufous-breasted Wren, Turquoise Tanager, and both Trinidad and Violaceous Euphonias.

After about 200 m the path becomes a grassy trail leading into the forest. There is a large concrete spillway and sluice on the right. After another 200 m the trail descends to the tributary, which is crossed along the top of a small dam wall and then ascends into the forest once more. Watch (and listen) for Little Tinamou, trogons, Gray-throated Leaftosser, White-bellied Antbird, Black-faced Antthrush, White-bearded and Golden-headed Manakins, and Red-crowned Ant-Tanager. The trail (which narrows farther on) is about 3 km long and is often blocked by fallen trees.

To leave the reservoir simply retrace your route 5.1 km back to Valencia Road, stopping by the forest alongside the Quare River. Turn right at Valencia Road, which rejoins the Eastern Main Road after 0.65 km.

Aripo Livestock Station
20 km – Half a day

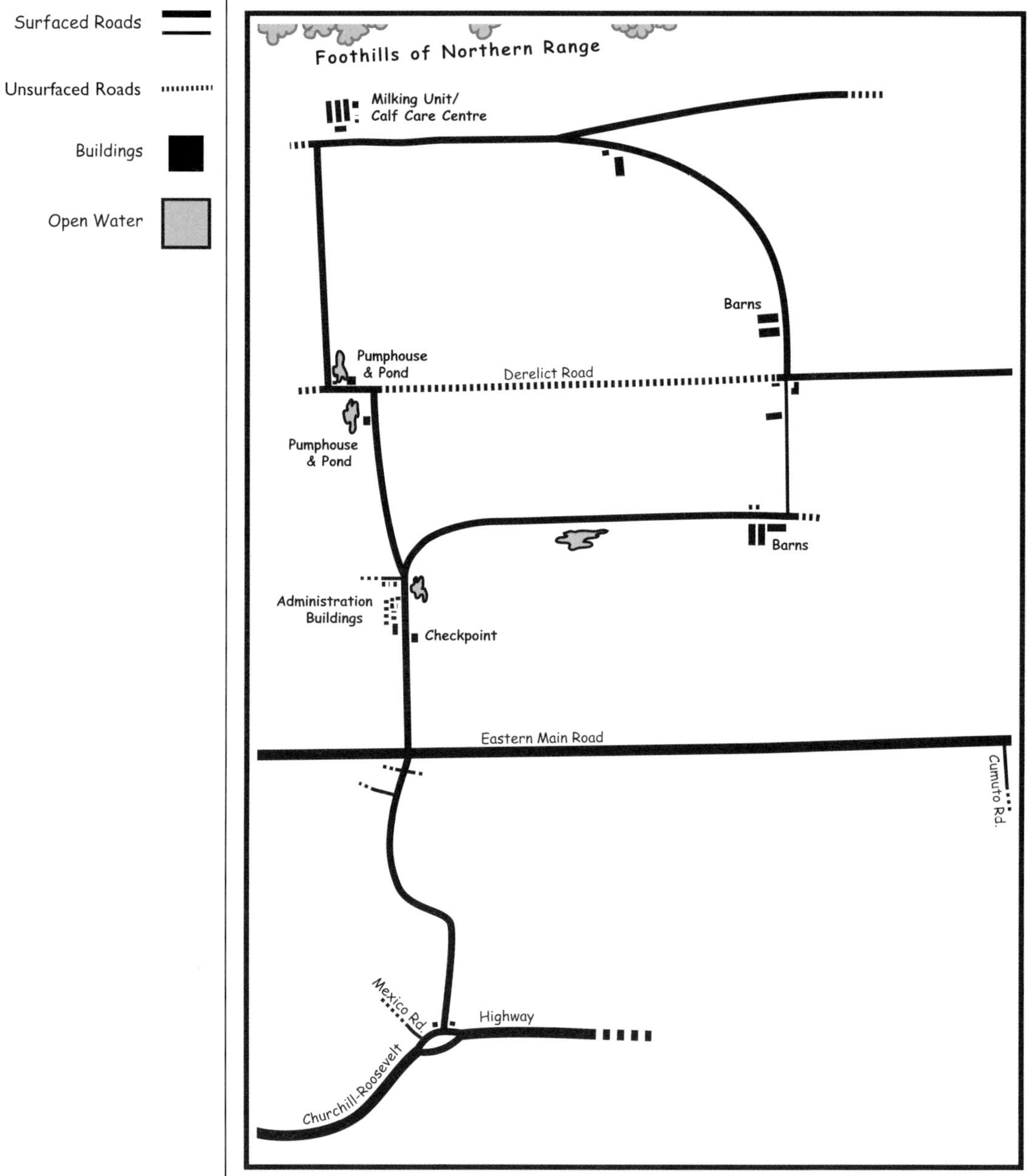

The Aripo Livestock Station is an active cattle ranch and dairy farm a few kilometres east of Arima in the foothills of the Northern Range. The areas of primary interest are the extensive wet pastures, brushy hedgerows, and a few huge isolated trees that serve as strategic vantage points for raptors, especially Savanna Hawks, Merlins, and Peregrine Falcons.

To reach the Aripo Livestock Station from the Arima area follow the

Churchill-Roosevelt Highway east of Tumpuna Road (junction with traffic lights) and continue for 2 km, ignoring a left turn onto Pinto Road after the first kilometre. Continue along this highway ignoring a left turn after 400 m (Demerara Road) and another after 2.1 km (Mexico Road), but take the next left turn, about 100 m farther on. Two small stone pillars mark the entrance to this road, which should be followed for 1.2 km, ignoring any side roads, to an intersection with the Eastern Main Road. Straight ahead is the entrance to the Aripo Livestock Station. Look for Savanna Hawks and Green-rumped Parrotlets in the trees that border the road to the station. Flying overhead or feeding in the fields will be White-headed Marsh-Tyrant, Pied Water-Tyrant, Blue-black Grassquit, Carib Grackle, Shiny and Giant Cowbirds, and Red-breasted and Yellow-hooded Blackbirds.

Drive 200 m to the gate/checkpoint and explain that you are birdwatching. Park across from a series of buildings on your left. Survey the fields and the small watering hole just inside the wire fence. Watch for Southern Lapwing, Wattled Jacana, and, in the proper season, Collared and Semipalmated Plovers, Solitary, Spotted, Least, and Pectoral Sandpipers, Ruddy Turnstone, and Short-billed Dowitcher.

From this point it is most productive to drive short distances, stopping regularly to scan the fields, trees, and open sky along the way. As shown on the map, it is possible to complete a circuit of the Aripo Livestock Station. The clockwise route is described here.

Proceed for about 50 m to where the road bends to the right. There are two side roads to the left. Ignore the first and turn onto the second. Continue for about 500 m, birding as you go. Common Black-Hawk, Savanna Hawk, Zone-tailed Hawk, Yellow-headed Caracara, Merlin, and Peregrine Falcon hunt over the upper fields. Look closely for the yellow-banded napes of the Turkey Vultures roosting on the ground in the upper fields, a characteristic of the Venezuelan race of the Turkey Vulture. There is a small pump house and pond on the left. Watch for Purple Gallinule, Common Moorhen, and Yellow-chinned Spinetail, as well as Spectacled Caiman that may be cruising about. Straight ahead is a T-junction. The derelict trail to the right is a useful one to explore on foot as you scan the fields, shrubs, and fence lines for Pinnated Bittern, Great Blue and Cocoi Herons, Stripe-breasted Spinetail, Gray-breasted Martin, Southern Rough-winged and Barn Swallows, and Ruddy-breasted Seedeater.

Proceed back to your car and continue by turning left. After 150 m turn right alongside another pond and pump house – a good spot to see Pied Water-Tyrant and White-headed Marsh-Tyrant. Continue along this road for 0.8 km. There may be barbed-wire partitions across the road. Unhitch these and pass through, making sure you replace them. At the T-junction turn right, and after about 100 m there are a series of buildings (a milking station/calf care centre) on your left. Stop and cautiously approach this area on foot, paying particular attention to all ground-doves because this is a reliable area for the not-so-common Common Ground-Dove.

The next stretch of road is usually good for Giant Cowbird, which is often seen walking in the pastures on the left. After about 0.7 km the road forks. The left branch is a 1-km cul-de-sac with lots of roadside shrubbery. Check this area for Striped Cuckoo and the elusive Bran-colored Flycatcher. Drive back to the fork and turn right as the road curves right. After about 1 km there are two large barns on the right. Keep scanning the fields and trees as you proceed – uncommon species such as Dark-billed Cuckoo and Gray Kingbird have been recorded at the Aripo Livestock Station. About 250 m beyond the barns is a four-way junction. To the right is the derelict road explored earlier on foot from the other end. To the left a road proceeds about 1 km through more scrub and pasture. Good unobstructed views make this another promising spot to search for Striped Cuckoo (often seen perched on fence posts or shrubs).

Going straight ahead at this junction leads to a T-junction after about 400 m. Turn right and go past more barns on the left and muddy cattle-holding areas also on the left where various shorebirds can be observed. About 1 km farther is the beginning of the circuit. Continue past the checkpoint and retrace the route back to the Eastern Main Road.

Arena Forest
40 km – 1 day

The Arena Forest is a lowland rainforest located in north-central Trinidad. The word 'arena' refers to sand, the predominant soil type locally. The terrain averages 50 m above sea level and is gently rolling, in sharp contrast to the nearly vertical terrain of the Northern Range. The Arena Reservoir is located here. A permit is required to enter the vicinity of the dam. It must be obtained in advance from the Water and Sewerage Authority (WASA), Farm Road, St. Joseph, Trinidad, for TT$5.50. Species seen include raptors, trogons, woodpeckers, woodcreepers, manakins, tanagers, hermits, and many insectivorous and ant-associating species.

To reach the Arena Forest follow the Churchill-Roosevelt Highway east to Tumpuna Road (south of Arima). Look for the blue-and-white 'Arena Dam' sign on the left just before the traffic lights.

Once on Tumpuna Road bear right at 2.1 km where the road curves sharply, cross a steel bridge over the Caroni River at 3.1 km, and enter the village of San Rafael. After 4.2 km there is a statue of St. Rafael on the right and a church on the left. Turn left at the T-junction and continue for 1.1 km to a junction where the road curves to the right. Stay on the right. This becomes Arena Road, which leads to the Arena Forest and Dam. The road deteriorates on reaching some small chicken farms 0.9 km farther on, where the forest begins.

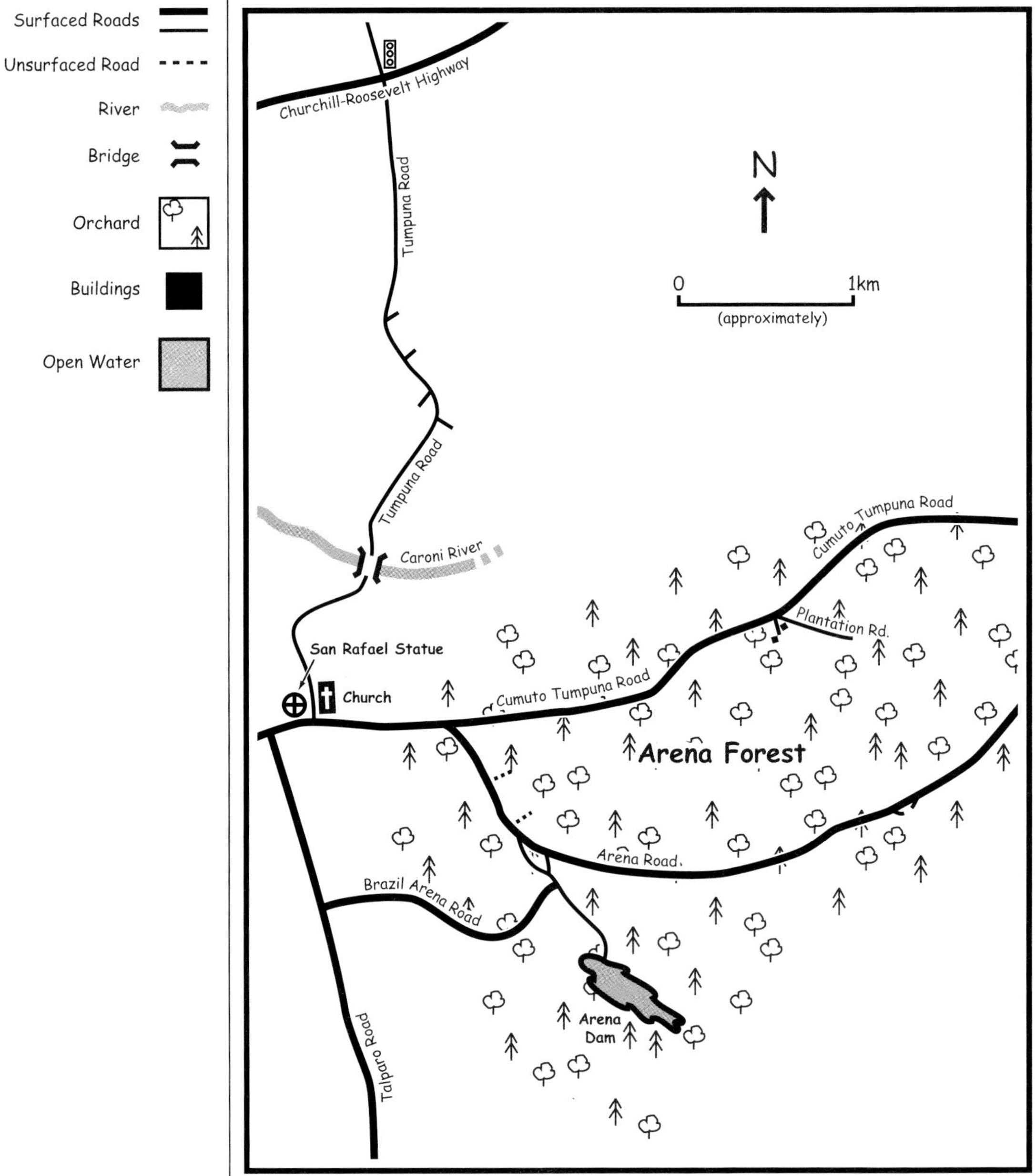

Drive 0.5 km, park, and commence birding along a trail leading left into the forest, looking and listening for Little Tinamou, Plumbeous Kite, White Hawk, Blue-headed Parrot, Little and Striped Cuckoos, hermits and other hummingbirds, trogons, Blue-crowned Motmot, Rufous-tailed Jacamar, Channel-billed Toucan, Red-rumped, Golden-olive, and Lineated Woodpeckers, Pale-breasted and Stripe-breasted Spinetails, woodcreepers, Great, Black-crested, and Barred Antshrikes, White-bellied Antbird, Black-tailed Tityra, and White-bearded and Golden-headed Manakins.

Drive another 0.5 km until you see a trail on the left that leads into an

Arena Reservoir

open, scrubby area. Walk in and scan for raptors, especially soaring groups of Plumbeous Kites. Look in the brush and grass for Sooty Grassquit and Ruddy-breasted and Yellow-bellied Seedeaters. After 100 m the road forks. Stay to the right and proceed 0.3 km to a T-junction. Turn right and remember this spot for the return trip. Ignore a side road on the right after 100 m and another at 0.8 km. At the top of an uphill stretch (0.2 km), permits must be shown to the WASA officer before proceeding to the dam.

Bear right and park in the car park next to the warden's house. On the left are the reservoir and the Central Range in the distance; on the right is the Northern Range. The dam wall is straight ahead. Walk across the dam scanning both the reservoir and spillway for Least and Pied-billed Grebes, Neotropic Cormorant, Anhinga, Cocoi and Little Blue Herons, Great Egret, Fulvous and Black-bellied Whistling-Ducks, Osprey, Swallow-tailed and Plumbeous Kites, Common Black-Hawk, Gray Hawk, Yellow-headed Caracara, Peregrine Falcon, Common Moorhen, Black-bellied (Gray) Plover, Laughing Gull, Large-billed and Yellow-billed Terns, Black Skimmer, and White-winged, Southern Rough-winged, and Barn Swallows. At the end of the dam retrace your steps back to your car.

Drive back to the entrance gate and bear right towards the reservoir. Park next to a small picnic area on the right. There are public lavatories here. Walk along the narrow footpath to a concrete shelter at the end of the peninsula. Bird as you walk, paying attention to the nearby trees, grasses, and water's edge and distant shoreline for the species mentioned above as well as for Green and Striated Herons, Green Kingfisher, Yellow-bellied Elaenia, Pied Water-Tyrant, White-headed Marsh-Tyrant, Piratic Flycatcher, and Masked Yellowthroat. You should be able to find at least one of the many

Spectacled Caimans that often bask along the water's edge.

From the car park the road continues straight ahead and then turns left, then right, into a final picnic area. From the final picnic area a narrow path (on the left as you leave the picnic area) provides access to a site from which reservoir can be scanned from a different angle.

Exit through the gate and drive 1.0 km, ignoring the two left turns, until you reach the junction remembered from your way in. Turn right and proceed 100 m to a T-junction. Turn right onto a narrow road that runs through the heart of the Arena Forest. Several productive forest trails begin on either side of the road. At a Y-junction after 1.8 km, turn left and proceed 3.0 km to the Cumuto-Tumpuna Road.

Turn left at this junction. If you wish to bird farther look for the Arena Forest Reserve Rest House on the left after 2.2 km. Prominent Caribbean pines border the entranceway. Park on the entrance driveway and search the evergreens for roosting Tropical Screech-Owl and Ferruginous Pygmy-Owl. Beyond a small shelter house beside the road is the beginning of Plantation Road, which runs deep into the forest. To retrace your steps, turn left, back onto the Cumuto-Tumpuna Road, and proceed 2.5 km to a T-junction that may look familiar – to the left is Arena Road. Turn right and proceed 1.1 km back to the junction in San Rafael village (with the church and statue). Turn right and retrace your route 4.2 km back to the Churchill-Roosevelt Highway.

Mount St. Benedict
15 km – Half a day

The Mount St. Benedict area comprises a number of different habitats including open grassland, tropical pine forest, and semi-cultivated, primary, and secondary forests. There are numerous good vantage points, making this a fine area for observing raptors and swifts. Birds found here include Little Tinamou, White, Common Black-, Short-tailed, White-tailed, and Zone-tailed Hawks, Ornate Hawk-Eagle, Yellow-headed Caracara, Bat Falcon, Pale-vented Pigeon, White-tipped Dove, Squirrel and Striped Cuckoos, Tropical Screech-Owl, Spectacled Owl, Ferruginous Pygmy-Owl, Common Pauraque, Rufous Nightjar, Chestnut-collared and Lesser Swallow-tailed Swifts, Fork-tailed Palm-Swift, Green Hermit, Tufted Coquette, Blue-tailed Emerald, Streaked Xenops, Southern Beardless-Tyrannulet, Bright-rumped Attila, Streaked, Boat-billed, and Piratic Flycatchers, Golden-headed Manakin, Rufous-browed Peppershrike, Hepatic Tanager, Grayish Saltator, Sooty Grassquit, Shiny and Giant Cowbirds, and Yellow Oriole.

To reach Mount St. Benedict, follow the Eastern Main Road to

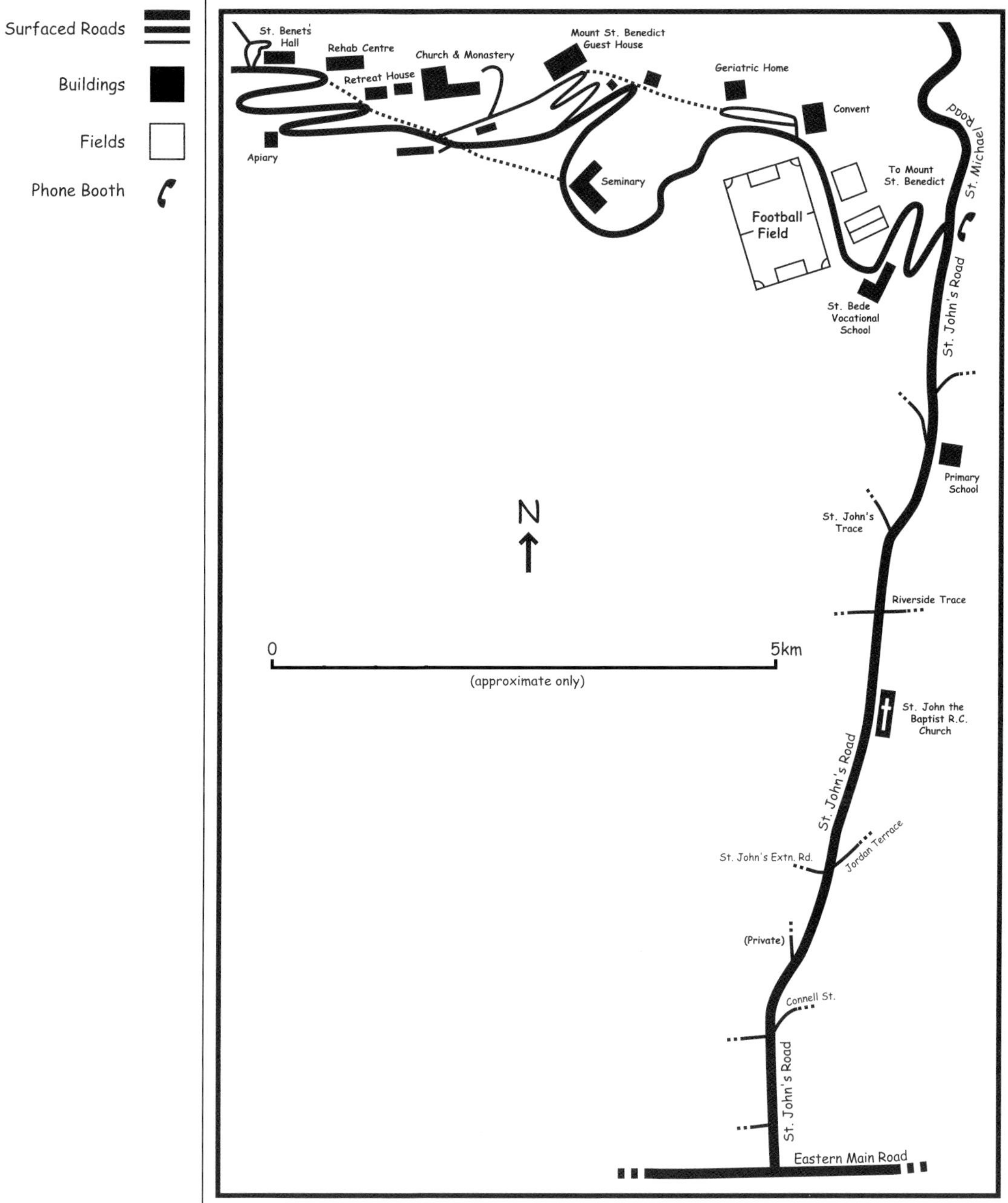

Tunapuna, which is located about 20 km east of Port-of-Spain. Turn north towards the mountains from the Eastern Main Road in Tunapuna, onto St. John's Road, which is just east of the Scarlet Ibis Hotel.

Follow St. John's Road uphill, ignoring the numerous side roads and alleys. Proceed past St. John the Baptist Roman Catholic Church on the right at 1.0 km. Bear right at the Y-junction after 1.2 km. Continue past a primary school on the right at 1.4 km. Proceed uphill to the junction of St. John's Road, St. Micheals Road, and the entrance drive to the Mount St. Benedict complex at 1.9 km.

Turn very sharply left and head uphill. During daylight hours there is no need to stop at the security gate; after dark, stop and tell the guard that you are a birder. Two switchbacks beyond the gate, you will drive past the St. Bede Vocational School and a playing field, both on the left. Follow the road along the perimeter of this field. Proceed past the Regional Seminary on the right, and after the next left switchback, at 1.0 km, look for a small sign saying 'PAX Guest House' on an inconspicuous right switchback from the main road. This takes you through three switchbacks to the car park of the guest house. If you miss this turning, about 250 m farther on make a hairpin right turn at a sign on the left that reads 'Apiary, Retreat House, Rehab Centre'. This road will also take you to the guest house. Turn hard right, stay on the right past a Y-junction, and proceed 200 m over a slight hill and around a bend to the guest house on the left.

Gerard Ramsawak, the manager of the Pax Guest House, is a keen birder who will offer advice and answer questions. The car park is a great area for viewing swifts, raptors, parrots, and parrotlets. Nearby vegetation may yield White-tailed and Violaceous Trogons, Great and Barred Antshrikes, Ochre-lored Flatbill, Long-billed Gnatwren, Red-eyed ('Chivi') Vireo, and Golden-fronted Greenlet as well as the usual species.

To bird the higher reaches of Mount St. Benedict return to the 'Apiary, Retreat House, Rehab Centre' sign just up the hill from the guesthouse. Watch for Gray-headed Kite and Zone-tailed Hawk, which patrol the slopes below, and for Blue-tailed Emerald, Long-billed Starthroat, and Streaked Flycatcher, which nest in this area. Continue your ascent by proceeding straight ahead. Make a right hairpin turn (the apiary is located on this corner), a left hairpin turn (avoiding the archway that leads to the retreat house and monastery), another right hairpin turn (overlooking the apiary), and a final left hairpin turn at 0.4 km in front of the Rehab Centre, then up the hill to the St. Benet's Hall car park. The road continues a short distance before ending slightly farther uphill next to a long building, a scout house/craft shop, which marks the beginning of the western start to the trail. If starting from the guest house, it would be best to walk the outlined route, scanning the fields, trees, and shrubbery along the way and looking overhead for raptors, even though a more direct footpath can be followed from the road in front of the Guest House to St. Benet's Hall.

From the St. Benet's Hall car park the forest can be explored on foot. Watch for Little Tinamou, White-tipped Dove, woodpeckers, woodcreepers, Ochre-bellied Flycatcher, Golden-headed Manakin, all three hermits, White-tailed and Violaceous Trogons, and the usual forest species.

The trail is actually a circuit loop and can thus be walked from either the east or the west. The western approach is very steep; therefore it is easier to tackle this trail from the eastern side, which begins in the parking lot proper, so as to make the more difficult section downhill rather than uphill. To start, look for the trailhead that begins below two water tanks at the western end of the concrete wall.

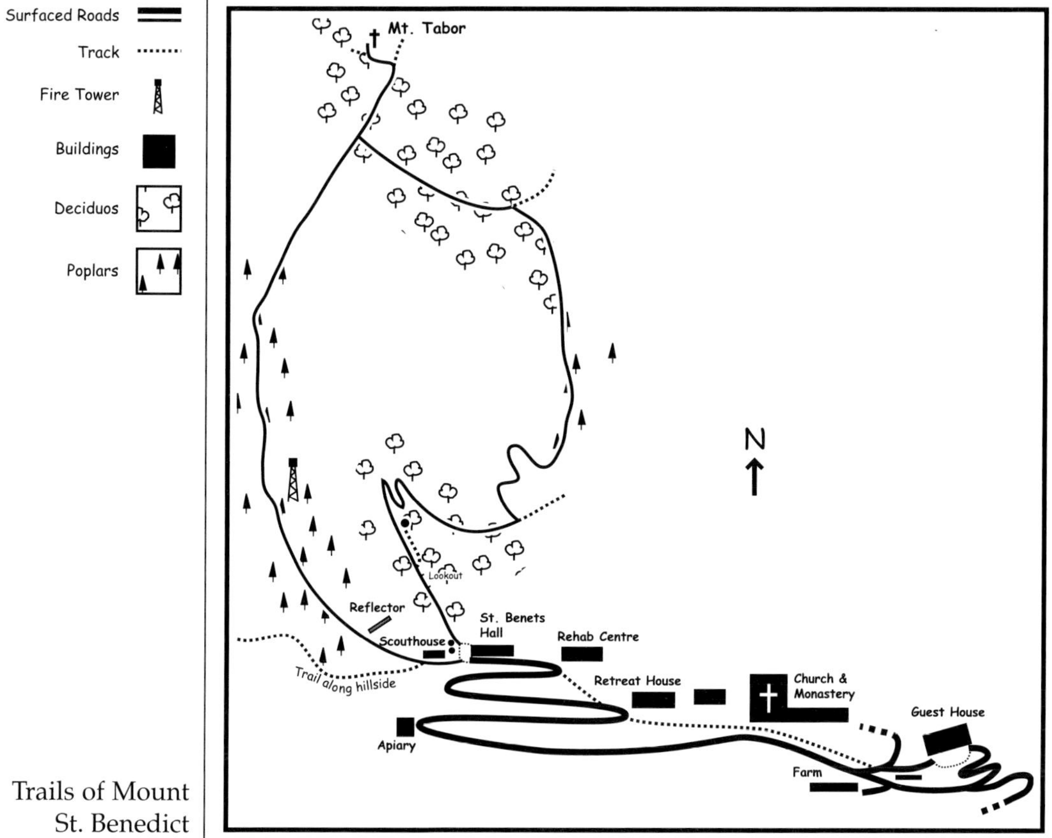

Trails of Mount St. Benedict

Proceed along this trail, sharply downhill at first, birding as you go. After 170 m you will arrive at a 30-metre-long steel footbridge clinging like a philodendron to the cliffside. The bridge affords an excellent view across the forest canopy and over the ravine below to the opposite ridge. About 50 m beyond the bridge the trail forks. Keep to the left (uphill). After about 160 m the trail crosses the headwaters of a small stream. The trail then switchbacks to the right, then the left, then again to the right. Continue through the forest for about 250 m and look for an inconspicuous intersection. Take the sharp left here and follow this route for about 350 m into a more open area with tall grass, shrubs, and Caribbean pines. From here glimpses of the Plains of Caroni to the southeast can be seen. You are now on a side ridge that leads up to the main ridge. After about 360 m another inconspicuous junction is reached; turn left and after 280 m there is a prominent intersection that marks your arrival at the main ridge. Here there are two options:

(1) Those wishing to go still higher to bird Mount Tabor should turn right at this intersection and after 180 m make a sharp left. In another 60 m the weatherworn remains of the original monastery are visible on a small hillock to the right.

(2) Those wishing to proceed along the main ridge should turn left. After about 200 m an open area with many ferns and large quartz boulders is reached. Here there are spectacular views into the Maracas Valley (to the right) and down past the monastery complex onto the Plains of Caroni (to the left). This area is as good as it gets for raptor watching. Find some shade among the pines just ahead and scan the Maracas Valley for any birds of prey that may be rising on thermals. Patience usually pays off here. After about another 150 m there is a rustic bench. At this point the trail turns left and begins a sharp descent through a grove of Caribbean pines that were transplanted here from their native island of Hispaniola. Follow this steep section for 350 m until a fire tower is reached. Be careful, though – this trek may be quite slippery and tough on the knees if you slip. Those who are not afraid of heights might risk ignoring the warning signs and climb the fire tower to scan for raptors and swifts. A further 250 m below the tower is a large reflector board on the left. About 50 m beyond that is the long building (scout house/craft shop) just above the St. Benet's Hall parking lot, where this hike started.

Another productive trail begins at the western end of the scout house (under the large forestry sign). This trail skirts westward along the southern face of the hillside for several hundred metres through vegetation consisting of Caribbean pines, scrub, and tall grass. The area is good for Long-billed Starthroat, Golden-olive and Lineated Woodpeckers, Plain-brown and Cocoa Woodcreepers, Streaked Xenops, Rufous-breasted Wren, Sooty Grassquit, and the usual species.

Waller Field and Aripo Savannah 20 km – Half a day

The area southeast of Arima contains one of the last remaining natural savannahs in Trinidad. Birding can be very productive here throughout the year. Two areas of particular interest are Waller Field and the Aripo Savannah. Waller Field is an abandoned U.S. airbase dating back to World War II. It consists of two east-west runways serviced by a network of roads. Except for the sterile concrete runways and roads, the area is overgrown with trees, brush, and grass.

The Aripo Savannah is a seasonally wet prairie with a unique assemblage of trees and plants. It is possible to see 70 species of birds in the region including Gray-headed, Hook-billed, and Pearl Kites, Savanna Hawk, Gray Hawk, Yellow-headed Caracara, Red-bellied Macaw, Striped Cuckoo, Fork-tailed Palm-Swift, Ruby-topaz Hummingbird, Southern Beardless-Tyrannulet; Lesser Elaenia, Bran-colored and Sulphury Flycatchers, the extremely rare Moriche Oriole, and Yellow-rumped Cacique. It is also one of the most reliable spots in Trinidad for glimpsing the introduced Small Indian Mongoose.

To reach Waller Field drive east from Arima on the Churchill-Roosevelt Highway. Pass between two stone pillars that mark the beginning of a

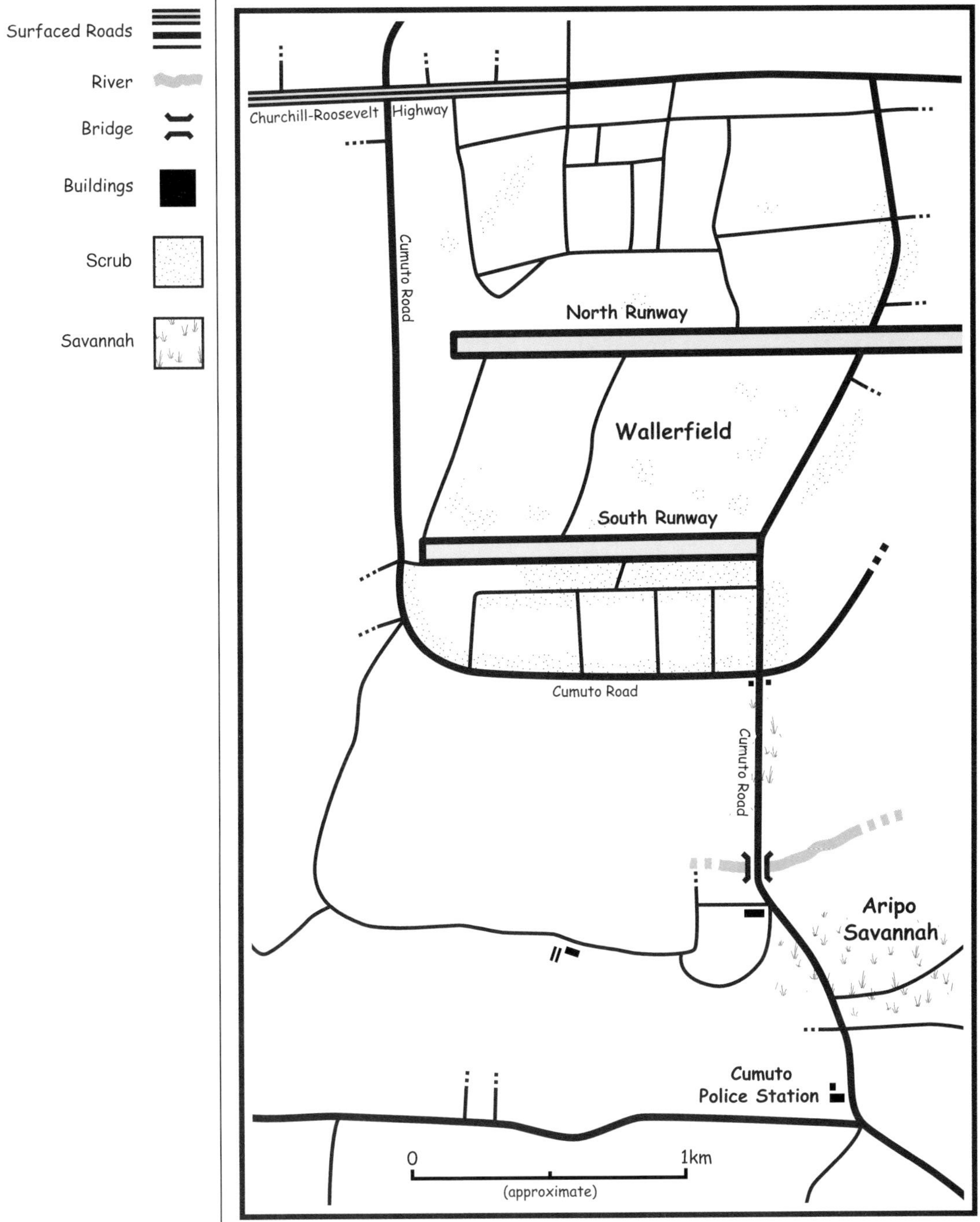

dual carriageway and the boundary of the former U.S. Army base called Fort Reid. Proceed for about 3.5 km to a crossroads. Turn right onto Cumuto Road.

Follow Cumuto Road as it veers sharply left after 2 km. At 3.7 km you will come to a sharp right-hand bend. Instead of following this bend, turn sharply left and park. This is the south entrance to Waller Field. There is a sign on the right saying 'Waller Field Racing Circuit du

Maurier Supercircuit Championship Competitors Entrance'. The cattle pasture behind this spot often holds Southern Lapwing, Wattled Jacana, and an occasional Savanna Hawk or Yellow-headed Caracara. Ahead is the south runway of Waller Field. The vegetation on either side of this short road is good for hermits and other kinds of hummingbirds, Blue-crowned Motmot, Rufous-tailed Jacamar, Black-crested and Barred Antshrikes, Bran-colored Flycatcher, White-winged and Southern Rough-winged Swallows, Blue Dacnis, and Turquoise and White-shouldered Tanagers.

Brushy margins of Waller Field

On the left about 50 m down this entrance road is an inconspicuous lane that connects with a network of roads south of the runway. Look here for Hook-billed Kite (rather sedentary), Squirrel Cuckoo, Black-throated Mango, Rufous-tailed Jacamar, Black-crested Antshrike, Ochre-lored Flatbill, and Black-tailed Tityra. After about 0.5 km a short connecting road on the right leads to the south runway.

Drive and walk as much of the Waller Field area as possible, exploring all of the areas accessible by road. Huge flocks of Black Vultures congregate on the runways, Zone-tailed Hawks mimic Turkey Vultures overhead, and at times as many as three Hook-billed Kites have been seen at once. In the rainy season flooded areas may attract such species as Southern Lapwing, Wattled Jacana, Yellow-chinned Spinetail, Pied Water-Tyrant, and White-headed Marsh-Tyrant. In the evening this is a good place to use a spotlight to look for the reflected eye-shine of Common Pauraques and White-tailed Nightjars and to try for the rare Nacunda Nighthawk on migration. After dark the still-warm concrete surfaces at Waller Field host scores of nightjars and Common Pauraques. Barn Owl and Common Potoo sometimes can be heard or seen overhead or in nearby bushes.

A note of caution. Waller Field is an isolated area away from main thoroughfares. It is wise not to attract too much attention. Stay close to the car at all times. Night birding here is advised only if part of a larger group.

Next return to the access road and continue south on Cumuto Road, immediately passing between a pair of old stone posts that used to mark the southern entrance to the airbase. Cross the muddy Aripo River via a steel bridge at 0.9 km and continue around a left bend. At 1.3 km turn sharply right so that a large derelict building is now on the left. Scan the fields, bushes, and treetops on the right for such species as Plumbeous Kite, Southern Lapwing, Rufous-tailed Jacamar, Wilson's Snipe, and Giant Cowbird. Proceed 0.4 km and turn left at a T-junction. Water Buffalo graze in the nearby field. Proceed another 0.3 km carefully scanning the pasture, distant trees, and sky for Savanna Hawk, Yellow-headed Caracara, Peregrine Falcon, Green-rumped Parrotlet, Fork-tailed Palm-Swift, Black-crested Antshrike, and Fork-tailed Flycatcher.

Follow the road as it veers right and proceed 0.4 km past some small chicken farms on the left. The open grassy area on the left is a good place to stop and search for Striped Cuckoo, Ruby-topaz Hummingbird, White-flanked Antwren, and Sulphury Flycatcher. It is possible to proceed farther along this road, birding along the way, but it gradually becomes more closed in.

Retrace your route to Cumuto Road and turn right (not a hard right onto the small lane). Moriche palms, actually a type of palmetto, are soon visible on both sides of the road. At 0.5 km look on the left for an obscure track that penetrates deep into the Aripo Savannah. The savannah itself is a fragile community that the government has designated a restricted area, so do not venture off the lane. Drive into

Yellow-rumped Cacique nests in Cumuto village

the savannah or walk if the path is muddy, birding the shrubs and palms for Red-bellied Macaw, Fork-tailed Palm-Swift, Ruby-topaz Hummingbird, Lesser Elaenia, Sulphury Flycatcher, and Moriche Oriole.

Retrace your route to Cumuto Road, turn left, drive 0.6 km, turn right, and park at the Cumuto Police Station/Health Centre with its long-standing colony of Yellow-rumped Caciques ('kuh-SEEKS') in the Caribbean pines. During the week the toilets at the Health Centre are available on request. At LC's store across the road, Lenore Noray and her family offer shade, cold drinks, tasty food, and a friendly reception to birders.

To return to the Churchill-Roosevelt Highway retrace Cumuto Road for 6 km.

WETLANDS

Caroni ('Care-uh-knee') Swamp
10 km – Half a day

Mangrove-bordered canal through Caroni Swamp

An evening excursion into the Caroni Swamp is considered one of the highlights of a visit to Trinidad. A spectacular event is the flight back to their roosts of egrets, herons, and Scarlet Ibis that begins between 4 and 5 p.m. and continues until dark. Except during the breeding season these large waterbirds roost communally on mangrove hammocks, with the Scarlet Ibis – the star of the show – forming the upper tier and the herons and egrets settling mostly in the lower branches.

The Caroni Swamp Bird Sanctuary is located 20 minutes south of Port-of-Spain and extends approximately 25 km along the Gulf of Paria. The swamp proper is approximately 65 km^2 in size and was created by land depression south of the Northern Range. The government has set aside about 182 ha (450 acres) within the swamp as the Caroni Swamp Bird Sanctuary. The entrance is accessible from the Uriah Butler Highway, which forms the eastern boundary of the Caroni Swamp.

The entire Caroni area is tidal, with a difference of about 1 m between high and low tides. Because of the cyclical nature of rainfall in the tropics, the area is influenced by the intrusion of copious quantities of fresh water from the interior, particularly during the wet season, which normally extends from May to December. The mixture of fresh and salt water creates the brackish water condition that is crucial to the diversity of the Caroni's flora and fauna.

The vegetation within the Caroni Swamp can be described in general as 'mangrove'. The term is not a taxonomic classification and thus may be used correctly to cover all vegetation found in the intertidal zone or substrate. However, the Caroni is dominated by a single species of rhyzophore, which is commonly called 'red mangrove'. This type of mangrove ecosystem is the reason why animal life abounds in the Caroni. It serves as the foundation of a complex food chain that attracts and supports a wide variety of birds, more than 185 species of which have been recorded from the Caroni Swamp.

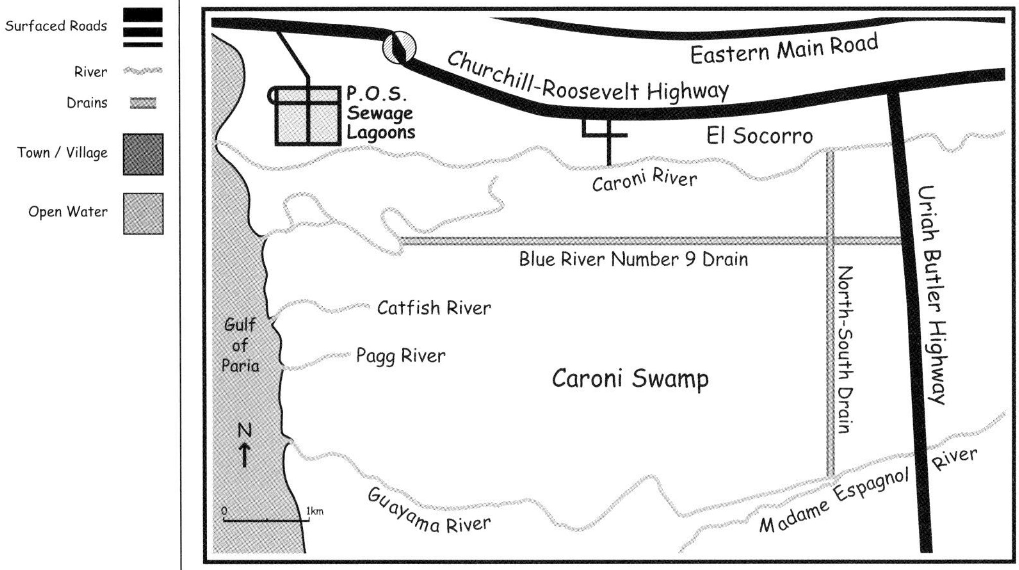

To reach the Caroni Swamp Bird Sanctuary begin at the junction of the Churchill-Roosevelt and Uriah Butler Highways. Drive south 5 km to a sign that says 'Caroni Bird Sanctuary'. Follow the signs, crossing the overpass and descending on a side road that runs along the right side of the highway. Turn left and park near the docks on either side of the channel, the technical name of which is Blue River Drain Number 9. If you have time, drive west along the south shore of the river to the modern visitor centre, which has toilets and information about the Caroni Swamp.

Before embarking on a boat trip into the Caroni Swamp, verify that the boatman is familiar with the speciality birds of the mangroves and will endeavor to find not only the Scarlet Ibis but also the rarer species that there is little chance of seeing elsewhere: Boat-billed Heron (nocturnal), Common Potoo (nocturnal), Mangrove, Little, and Dark-billed Cuckoos, Greater Ani, Green-throated Mango, Straight-billed Woodcreeper, Bicolored Conebill, and Red-capped Cardinal. It is best to hire a professional birding guide who will pilot the boat to prime areas in response to specific requests from birders.

Spend some time birding the area on foot since many species of birds found only with great difficulty elsewhere in Trinidad are found easily in the mangroves bordering the channels. Apart from those already mentioned look for Pied-billed Grebe, Neotropic Cormorant, Anhinga, Pinnated, Stripe-backed, and Least Bitterns, all of the common egrets and herons, Roseate Spoonbill, Fulvous and Black-bellied Whistling-Ducks, Masked Duck, Long-winged Harrier (rare), Yellow-headed Caracara, Merlin, Peregrine Falcon, Aplomado Falcon (rare), Sora,

Purple Gallinule, Common Moorhen, Limpkin, a wide variety of shorebirds, Large-billed Tern, Yellow-chinned Spinetail, Fork-tailed Flycatcher, and migrant North American warblers, primarily American Redstart and Northern Waterthrush.

If time permits, walk along the south side of the channel. At the west end of the boat dock area, a trail leads to the left between a freshwater marsh on the left and a brackish marsh on the right, both good for rails, crakes, and bitterns. The road leads straight ahead along the river to the edge of a shallow lake, good for ducks, shorebirds, herons, egrets, terns, and Osprey. Watch for small flocks of chattering Green-rumped Parrotlets darting through the riverside mangroves. Birders often find resting Amazon (Cook's) Treeboas *(Corallus hortulanus cookii)* balled up among the outermost twigs of the mangroves. This species is not docile and can attain a length of two metres, so look but do not touch.

While birding along the channel, whether on foot or in a boat, scan the water's surface occasionally for Four-eyed Fish *(Anableps sp.)*. These unique creatures actually have only two eyes, divided into upper and lower halves, each half with its own focal length, endowing *Anableps* with simultaneous dual vision. Schools of these fish glide smoothly along just beneath the surface, the upper pupils protruding above water scanning for predators, the lower pupils scanning the aquatic world for morsels on which to feed.

Caroni Rice Fields
10 km – Half a day

The Caroni rice fields were created in the last decade by the Ministry of Forestry and Agriculture. The plains that stretch almost from Piarco International Airport to the Caroni Swamp were formerly flooded annually, which precluded traditional farming or cattle grazing. The vast quilt of rice fields that now occupy the western Caroni plains can be seen from aircraft on approach to the airport. A network of dikes, many with driveable tracks, makes it possible for birders to venture into most areas of the rice fields. At any given time some of the fields are newly ploughed, some are covered by newly sprouted rice, some have knee-high rice, and some have rice ready to be harvested.

The number of species of birds observed in the Caroni rice fields is awesome, as is the number of rare species: Stripe-backed Bittern, Grey Heron, Glossy Ibis, Roseate Spoonbill, Jabiru, all three species of whistling-ducks, Comb Duck, White-cheeked Pintail, Northern Pintail, Snail Kite, Long-winged Harrier, Aplomado Falcon, Yellow-breasted Crake, Azure Gallinule, Limpkin, Upland Sandpiper, Black-tailed Godwit, Buff-breasted Sandpiper, Ruff, Wilson's Phalarope, Short-eared Owl, Ringed Kingfisher, Masked Yellowthroat, Bobolink, and Common Waxbill.

To reach the Caroni rice fields start at the KFC at Valsayn. Turn right if you are coming from Port-of-Spain or left if you are coming from the east. Drive 2.4 km and cross the steel bridge over the Caroni River. At 4.6 km turn right just past the Caroni Police Station on the right. Watch for flocks of birds flushing up as raptors cruise over the fields.

There is another entrance to the Caroni rice fields near the Caroni Swamp boat docks. Note that from time to time this entrance is closed. To reach this entrance, follow the directions to the Caroni Swamp boat docks but instead of leaving the Uriah Butler Highway and crossing over the overpass, enter the rice fields via the track on the left about 6 m beyond the eastern end of the overpass.

El Socorro Marsh
10 km – Half a day

Channelling of the Caroni River began draining El Socorro Marsh in the late 1980's, driving out resident and migrant species of birds. In past years, sugar cane and vegetable fields planted around the perimeter of the marsh were flooded periodically because of siltation of the Caroni River, a result of clear-cutting and poor land management in the upstream watershed. It is hoped that conservationists will step in before the marsh is drained completely.

El Socorro Marsh has traditionally been one of the best places in Trinidad to observe a wide variety of unusual and elusive species such as White-faced Whistling-Duck, Masked Duck, Ash-throated and Yellow-breasted Crakes, Rufous-necked Wood-Rail, Sora, Spotted Rail, and the tiny Crested Doradito as well as more common species such as Yellow-chinned Spinetail. The marsh is located north of the Caroni River about 6 km east of Port-of-Spain. Unlike the brackish Caroni Swamp just to the south, El Socorro is a freshwater marsh with vegetation consisting of rushes, sedges, and a few woody shrubs. It is possible to bird the marsh from adjacent roads and lanes but to see the greatest number of species, explore the marsh itself. Watch carefully for protectively camouflaged bitterns that 'freeze' when threatened and for rails that skulk away almost unseen.

To reach El Socorro Marsh proceed west (toward Port-of-Spain) on the Churchill-Roosevelt Highway from the traffic lights at the Uriah Butler Highway. In 0.9 km, after passing a vast expanse of vegetable fields on the left, turn left onto Chootoo Road into a suburban housing area. Follow the road as it bends sharply to the right. Turn left at the sign for Zolaz exhausts on the left and follow the road a short distance to its end at the marsh. Park and walk out along a dike into the marsh, looking and listening for Large-billed Tern, Yellow-chinned Spinetail, Ruddy-breasted Seedeater, and Yellow-hooded and Red-breasted Blackbirds. It is possible to walk east for about 1.5 km to the San Juan River, south 0.5 km to the Caroni River, and west about 1.5 km to the heavily forested eastern edge of the Beetham landfill and sewage lagoons. Watch for the kestrel-sized Pearl Kite on the overhead wires.

Those who prefer to drive rather than walk should continue west towards Port-of-Spain along the streets closest to the marsh, making jaunts to their southern ends at the edge of El Socorro Marsh. Smooth-billed Anis, which are common in the brush, produce many 'mystery' avian sounds. Still other calls are produced by a number of different species of skulking marsh birds such as Rufous-necked Wood-Rail and Ash-throated Crake (both very rare), Yellow-breasted Crake and Sora (both abundant), and Spotted Rail (common). Playback is very effective when birding El Socorro Marsh, luring secretive species into view.

Nariva Swamp and Eastern Trinidad 80 km – 1 day

View from Plum Mitan Road

Birders visiting eastern Trinidad can explore a great variety of habitats in one all-day trip. The most productive birding spots are Brigand Hill, Plum Mitan Road, Manzanilla Beach/Mayaro Road, Bush Bush Creek, Nariva Swamp, the Melon Patch, and Point Radix.

Brigand Hill, the easternmost peak of the Central Range, provides an unforgettable panoramic view of most of eastern Trinidad as well as good birding in a region of semi-abandoned cocoa plantations and uninhabited rainforest. Before visiting Brigand Hill telephone or call in at the Forestry Division office in Sangre Grande to determine if the Brigand Hill lighthouse is open.

The Nariva Swamp occupies a vast triangular area along the eastern coast of Trinidad, extending from the village of Lower Manzanilla in the north almost to Point Radix in the south and as far inland as the Central Range. Freshwater habitat predominates except for brackish mangrove edges along the Nariva River and Bush Bush Creek. It provides excellent habitat for waders, rails, raptors, and at least one large aquatic mammal, the Manatee. A walk along the banks of Bush Bush Creek may yield Gray Hawk, White-tailed Kite, Limpkin, Green-throated Mango, Green Kingfisher, Black-crested Antshrike, and Silvered Antbird. In Nariva Swamp proper, waterbirds, both large and small, and Red-bellied Macaws are the main attractions. Finally, Point Radix offers spectacular views along the coconut-palm-fringed coast and close views of Black Vultures.

View from Maridale Road

To reach eastern Trinidad follow the Eastern Main Road past Valencia and then through Sangre Grande ('lots of blood'), the largest city in eastern Trinidad. Continue through the town of Upper Manzanilla, watching on both sides of the road for Yellow-rumped Caciques in flight.

Cross the bridge over the L'Ebranche River. At a wide T-junction turn right onto Plum Mitan Road. Drive 2.8 km, bearing left at a Y-junction to remain on Plum Mitan Road. A 4-wheel drive vehicle is useful here. At 3.2 km take a hard left onto a 0.8 km incredibly steep but well-paved road that climbs to the lighthouse atop Brigand Hill. Park when the road becomes too steep for traction and walk the rest of the way to the summit. Tell the guard at the top that you are birdwatching.

The Brigand Hill lighthouse is an automatic apparatus situated on top of open steel scaffolding. Walk up the incline to the lighthouse and climb the stairs to get unlimited views to the southeast of the entire Nariva Swamp area and beyond, a good view to the northeast past the satellite-tracking radio-telescope dish at Matura at the far eastern end of the Northern Range, and a reasonably good but partially tree-blocked view to the west along the spine of the Central Range. On the southeastern horizon it may be possible to see a fiery column of burning gas jetting from Texaco's offshore oil platform 70 km (43.5 miles) out in the Atlantic and from an even more distant jet in the Galeota oilfield. A large flock of Black Vultures can usually be seen at the base of the lighthouse and can be approached closely. You can also view several species of swifts advantageously from the tower.

View from La Victoria Road

On the way back down to Plum Mitan Road, stop at each switchback to bird. Look among the abundant heliconias for eight species of hummingbirds, including all three species of hermits, and overhead for trogons, elaenias, flycatchers, tanagers, euphonias, and others. This area is good for birds because of the tremendous numbers of insects attracted to the lighthouse at night.

Back on Plum Mitan Road drive 1.3 km and turn left onto the very rough Maridale Road, which follows the contours of Brigand Hill east along its base. Excellent birding is possible in this verdant rainforest. Canopy-loving species and hermits are particularly abundant. The uncommon Crimson-crested Woodpecker can also be found here. Formerly a cocoa plantation, this area has been allowed to return to a more diversified state. Explore for 1.5 km then turn around at the house at the end of the road and return to Plum Mitan Road.

Turn left on Plum Mitan Road, drive 0.2 km, and turn right onto the very rough La Victoria Road. The vegetation along this 12-km road is similar to that along Maridale Road but the area is slightly drier. The major difference between the two is that Maridale Road heads east, whereas La Victoria Road heads west. Thus it is best to bird along Maridale Road in the morning and along La Victoria Road in the afternoon.

To continue toward the ocean, return to Plum Mitan Road, turn left, and retrace your route 5.0 km back to its intersection with the Eastern Main Road. Cross the intersection and proceed east on Mayaro Road.

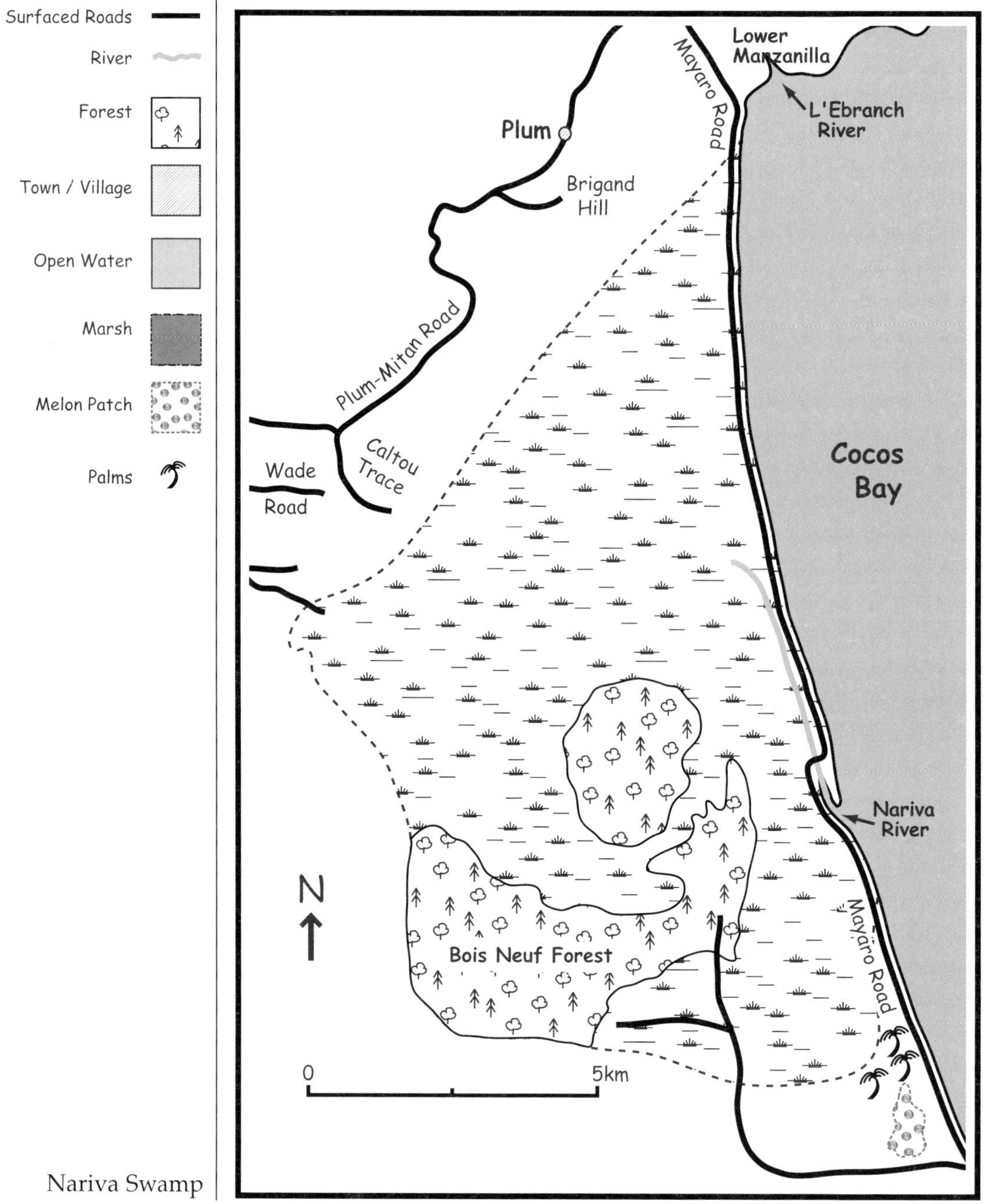

Nariva Swamp

(Consider this intersection to be 0.0 km in the following directions.) In about 3.6 km Mayaro Road drops steeply down to the seacoast along Cocos Bay. For the next 25 km it passes through vast numbers of coconut palms growing in a plantation on both sides of the road. The narrow strip of sand on which the road was built and on which the palms grow was a former barrier island.

Park along the beach wherever possible. Scan the surf for Brown Pelican, terns, and Leach's Storm-Petrel. This beach has produced most of Trinidad's petrel and shearwater records, mostly exhausted, ill, or dead birds. Watch the trunks of the coconut palms on both sides of the road for perched Common Black-Hawk, Gray and Savanna Hawks, and Yellow-headed Caracara.

Manzanilla Beach

On the right at 13.6 km are mangroves that provide a perfect habitat for a number of specialised bird species including Silvered Antbird and Bicolored Conebill. The mangrove edge is accessible for the next 9 km and should be checked well, although in wet weather a shallow depression between the road and the mangroves fills with rainwater and makes access difficult.

Gallery forest along Nariva River

At 15.0 km park at the Nariva River bridge. Walk onto the bridge and scan the mangroves for Anhingas, herons, and Limpkins. Groups of Red Howler Monkeys can sometimes be seen foraging in the trees at the western end of the bridge.

Red Mangroves near Bush Bush Creek

At 16.3 km park just before crossing the bridge over the stream, which is the legendary Bush Bush Creek mentioned by C. Brooke Worth (*A Naturalist in Trinidad*). Bird along the swamp edge as this is probably the best place in Trinidad to spot Silvered Antbird. Stop again at 16.9 km to scan the mangroves. Construction of concrete bridges over these waterways required substantial removal of mangroves, which reduced the likelihood of finding the target species. Just ahead the shoreline and transient sandbars at the mouth of the Nariva River sometimes hold resting shorebirds, terns, and Black Skimmers.

Manatee

Continue through this area scanning all the trees and fence posts for raptors, as this area has produced Gray-headed and Pearl Kites, Common Black-Hawk, Rufous Crab-Hawk, Gray Hawk, and Yellow-headed Caracara.

At 21.0 km note the field on the right, which contains 'islands' of royal palms. Between 5 p.m. and dark, Red-bellied Macaws fly in from neighbouring areas to roost in these trees. The macaws land in the vertical top fronds of the palms and slide slowly down into the foliage until they can no longer be seen. This spot is particularly good at dusk.

Red-bellied Macaws over Moriche palms

At 21.7 km turn right onto a one-lane paved road called Kernahan Trace, park, and bird the freshwater ponds on the right on foot. Continue along this road, which runs into a vast open area called the Melon Patch. This was a fabulously productive marsh until recently, yielding South America's first Western Reef-Egret in January 1986 and, more recently, has hosted a small colony of Azure Gallinules. In the early 1990s this section of the Nariva Swamp was ditched, burned, cultivated for rice production, then allowed to revert to a wilder state. Birds often seen include Purple Gallinule, Wattled Jacana, Yellow-hooded Blackbird, and a variety of raptors, such as Yellow-headed Caracara. The much smaller Tropical Kingbird often pursues the latter species.

The lesser roads and paths in this area lead into interior forests in which birds may be more abundant and tamer than in adjacent 'mainland' forests. Bands of Red Howler and Weeper Capuchin monkeys inhabit Bush Bush Forest and are best heard in the hours around dawn. During the Northern Hemisphere's winter, tens of thousands of Dickcissels often use the Melon Patch for feeding and roosting. Occasionally an Aplomado Falcon is seen with these flocks. At least one Pinnated Bittern can usually be found by persistent scanning. At dusk in the Melon Patch you may see Red-bellied Macaws flying over on their way to roost. Orange-winged Parrots will also be flying past in the distance. Other birds flying past include Great and Snowy Egrets and Little Blue Herons.

Point Radix

Return to the main road, turn right, continue along Mayaro Road, and cross a bridge over the Ortoire River. The fishing village of Ortoire is at the far end of the bridge. Take the first left out onto Point Radix. Bird to the gate of a private estate. Along the way pull off on the left at a conspicuous overlook. The view here is to the north and it is possible to see the thousands of coconut palms as well as several isolated rocks below, on which Brown Pelicans and an occasional tern rest.

If you wish to photograph Black Vultures at close range, take the only right turn on the way back to Mayaro Road. This lane leads to the mouth of the Ortoire River, where Black Vultures gather and feed among the fishermen. Green and Ringed Kingfishers may also be seen along this stretch of river, but the constant trade winds off the ocean make most landbirding difficult here.

Piarco Water Treatment Plant 1 km – 1 hour

Just across the runways from Piarco International Airport in north-central Trinidad is the Piarco Water Treatment Plant. A visit to this site can be a wonderful introduction to tropical birding. This little-known but highly worthwhile birding spot can yield a large number of interesting species. Currently it is only open during the week (Monday-Friday).

Piarco Ponds

To visit the site, obtain a permit in advance (see section on the Arena Forest for information on how to obtain the WASA permit).

To reach the Piarco Water Treatment Plant turn left on leaving the airport and follow the road along the perimeter fence to the south side of the airport. The treatment plant is the only facility on the right side of the road. Stop at the gate and give your permit to the guard, who will explain where it is permissible to go.

Park on the left just beyond the guardhouse. The guards always direct visitors towards two ponds that lie beyond the main buildings. The sterile-looking pond nearer to the airport is usually unproductive, whereas the farther pond is better because of its emergent aquatic vegetation. Species to be seen here include Anhinga, Southern Lapwing, Wattled Jacana, Yellow-billed and Large-billed Terns, Ruddy Ground-Dove, Pied Water-Tyrant, White-headed Marsh-Tyrant, Tropical Kingbird, Fork-tailed Flycatcher, Gray-breasted Martin, White-winged Swallow, Blue-black Grassquit, Red-breasted and Yellow-hooded Blackbirds, Carib Grackle, Shiny and Giant Cowbirds, Yellow Oriole, and at least one large reptile, the Spectacled Caiman. From November to March watch the skies for Peregrine Falcons and Merlins, which frequent the sugar cane plantations in the vicinity.

Pointe-à-Pierre Wildfowl Trust
160 km – Half a day

Molly Gaskin, Trinidad's most outspoken conservationist, established the Pointe-à-Pierre Wildfowl Trust in 1966 in the grounds of a petro-chemical refinery. Boasting two lakes and a profusion of native plants and trees, this exquisite compound near San Fernando on the central west coast of Trinidad is an interesting example of heavy industry and conservation working together.

The Trust is a national, non-profit making, volunteer organisation dedicated to and actively involved in the breeding of Trinidad's endangered species of waterfowl, cage birds, and indigenous species. The Trust also has initiated an environmental educational audiovisual programme shown in schools and to community groups throughout Trinidad and Tobago. The Trust property is a wetland habitat that also serves as an environmental education resource centre for scouts, guides, youth groups, and government ministries, providing training in the field and through action projects. Visiting hours are 8 a.m. – 5 p.m. Monday to Friday and 10 a.m. – 6 p.m. at weekends. Special arrangements are available for birders.

The main building – the Learning Centre – houses information dealing with living organisms and their habitats, a unique shell collection, and a small but comprehensive Amerindian museum.

Compounds and adjoining areas host myriad migrant waterfowl and shorebirds in the proper seasons. Other species can be seen throughout the year. Opportunities for close-up photography are excellent as the birds are accustomed to close approaches by humans. Species of particular interest include Neotropic Cormorant, Anhinga, Purple Gallinule, White-winged Becard, White-winged Swallow, Red-capped Cardinal, Sooty Grassquit, and Saffron Finch. Among the captive species are White-faced, Black-bellied, and Fulvous Whistling-Ducks, Canada Goose, Mute and Black-necked Swans, White-cheeked and Silver Pintail, Hooded Merganser, Laysan, Brazilian, and Blue-winged Teal, Common and Paradise Shelducks, Mandarin, Muscovy, Wood, Ring-necked, Rosybill, and Ruddy Ducks, and Red-crested Pochard.

The Trust should be contracted well in advance of visits to make reservations and secure detailed directions.

Pointe-à-Pierre Wildfowl Trust
c/o 38 La Reine Town House
Flagstaff Hill, Long Circular Road
St. James, Trinidad, West Indies
(868) 658-4210/4230/4200/ ext. 2512
or (868) 622-0238
or (868) 628-4145 (voice and fax)
e-mail: wildfowl.trust@petrotrin.com
web-site: http://www.petrotrin.com/wildfowltrustweb/papwft.htm

Port-of-Spain Sewage Lagoons

This is a dangerous area with a recent history of violent crime against birders at this site. It is probably best to avoid this location for the time being.

South Oropuche Swamp 160 km – Half a day

The South Oropuche Swamp attracts birders because of the presence of Spotted Tody-Flycatcher and Crested Doradito, along with Scarlet Ibis, a host of waterbirds, Greater Ani, Black-crested Antshrike, Bicolored Conebill, Red-capped Cardinal, and other mangrove-loving species.

To reach the South Oropuche Swamp from the north, drive to the south end of Uriah Butler Highway, follow the road as it loops to the left, then turn left at the T-junction. Proceed through one set of traffic lights to a roundabout. Continue straight through and pass through three traffic lights. After the third set stay in the right lane, passing a McDonald's/KFC/Pizza Hut complex on the left. Continue through one more traffic light to where the road divides. Follow the signs to 'M2 Ring Rd'. Turn left, drive a short distance, then turn right at the sign for 'La Fortune/Pluck' to drive through beautiful rolling fields of sugar cane. Look for a sign on the right that describes the South Oropuche Lagoon (Swamp). Turn right onto a narrow lane that serves as one entrance into the swamp. The road surface deteriorates but is driveable to the end.

Bird this road to its end, checking all the mangroves and open water for the species described above as well as for shorebirds, raptors (including Bat Falcon), and swallows. After exploring this road, return to the sign, turn right, drive for less than 2 km, and turn right at the T-junction. Follow this road for approximately 4 km, passing the Woodlands School on the right, to the Sudama Steps, which can be identified by a large sign on the right as you pass over a bridge with the concrete steps on the left. The track and trails along the south side of the stream, which is the Oropuche River, are excellent for birding. Look for Greater Ani, Spotted Tody-Flycatcher, Red-capped Cardinal, and the other species mentioned above.

Retrace your route to return to the North.

This trip can be combined with a visit to the following area for a very good day of birding.

Waterloo/Brickfield 120 km – Half a day

Situated along the Gulf of Paria a few kilometres south of the Caroni Swamp and just west of the village of Carapachaima, the Waterloo/Brickfield area is a waterbird haven. At low tide, extensive mudflats stretch almost as far as the eye can see to the north and south. The area is considered the premier gull-watching locality in Trinidad and Tobago and perhaps in the entire southern Caribbean. Most of the gulls seen are Laughing Gulls but also present at times have been Black-headed, Franklin's, Herring, Kelp, Lesser Black-backed, and Ring-billed. Gulls are best observed at low tide, when they congregate on the mudflats.

This is also a good place to find Gull-billed and Large-billed Terns. During migration, several dozen species of sandpipers, plovers, godwits and Whimbrel can be found here.

This trip can be combined with a visit to the South Oropuche Swamp for a very good day of birding.

To reach the Waterloo area, drive south from the Churchill-Roosevelt Highway junction in Valsayn on the Uriah Butler/Solomon Hochoy Highway. After 18.8 km exit left at the Freeport exit. At its end, turn right and continue 1.0 km to traffic lights on the Southern Main Road. Continue and proceed 2.0 km, passing the Waterloo High School on the left, to a T-junction. Turn left onto Orange Field Road and drive 3.9 km, turning right onto a paved road just beyond the fenced compound of the Caroni Research Station on the right.

Continue past the Carillion Waterloo Depot facilities on the left until you reach a sharp bend to the left and, after 0.9 km, a sharp bend to the right, but instead of turning right drive straight ahead. Continue 0.5 km to the paved car park of a fishing facility on the shore. This is the village of Brickfield. From the north side of the car park it is possible to walk or drive northward on a narrow track to get closer to the extensive mudflats (at low tide).

After scanning the mudflats at Brickfield, backtrack to Orange Field Road and turn right. Continue 0.5 km to a junction and continue straight across for 0.3 km to the car park of a Hindu temple built at the end of a causeway that juts into the Gulf of Paria. This area is known as Waterloo. Park at the northwestern end of the car park and walk northwards to a point of land overlooking an extensive mudflat (exposed during low tide) studded with old wooden pilings.

Search for gulls and terns, jaegers (skuas) harassing the gulls and terns, and for other waterbirds. Then drive south across the car park to an adjacent crematorium. Park at the southwestern end and scan for gulls and other waterbirds on the mudflats to the south. The main gull flock tends to shift unpredictably from one mudflat to another.

From Waterloo backtrack 0.3 km to the intersection and turn right, proceeding southward parallel to the shoreline. Along the next 1.5 km are several dirt tracks on the right that lead a short distance to the shore, where gulls and other waterbirds often congregate. These roads may be impassable when muddy.

After driving 2.1 km on the paved road, turn right onto Bay Street, the first paved road on entering the village of Orange Valley. Continue 0.6 km to a petrol station and fishing facilities on the shore. Search the mudflats for gulls and other waterbirds. Blue-and-white Swallows are sometimes seen hawking insects above the sugar cane fields in the area.

Another place where the nomadic gull flocks congregate is the San Fernando Public Fish Market. To reach it, retrace your route to the Uriah Butler Highway (called the Solomon Hochoy Highway from here south) and drive southwards until it ends in a left curve near San

Fernando. Turn left onto Tarouba Road and proceed 0.9 km to a set of traffic lights. Turn left at the lights and after 0.6 km turn right onto St. Joseph Road. Proceed 0.5 km and turn left onto London Street and then immediately right onto Circular Road. After 0.5 km veer left (still on Circular Road) and then left again onto Pointe-à-Pierre Road. After 0.5 km turn right onto St. James Street and proceed downhill 0.5 km until it ends. Turn left and immediately right onto a semicircle around Citibank. After 0.2 km turn left at the old Post Office into the Public Fish Market.

Usually a raft of gulls and pelicans is seen just offshore, waiting for fish offal discarded from the fish market. After scanning the area for gulls, drive back out of the market and turn left immediately after the fence, following the road out onto the pier, where more gulls may be seen to the right. Reverse your route to return to the highway.

Trincity Ponds (Home Construction Limited Sewage Treatment Works) 1 km – 1 hour

The Home Construction Limited Sewage Treatment Works (Trincity Ponds) near Trincity is a great place to obtain excellent views of a large number of water-loving species. The Trincity Ponds consist of six rectangular impoundments, two of which are covered with large masses of floating water hyacinths. Grass-covered dikes separate the ponds and provide easy access to all areas.

Trincity Ponds is the second most reliable place in the New World to see Little Egret, with Graeham Hall Swamp in Barbados being the most reliable. It is worth examining all white herons seen for one with bluish or greyish (not yellow) facial skin. Look for two long nuchal (head) plumes after January, which will confirm identification.

To reach the Trincity Ponds travel west along the Churchill-Roosevelt Highway from its junction with the road leading from the Piarco International Airport. Because this is a divided highway, it is not possible to access the ponds from the eastbound lanes. After 1.1 km, look for a row of Caribbean pines on the left – these trees form the eastern boundary of the site. At 1.3 km take the inconspicuous dirt road on the left, which leads into the Trincity Ponds.

If a guard is on duty, state that you are birdwatching. If no guard is present, drive through the metal gate, remembering to close it. Park near the Caribbean pines. The ponds are beyond the raised dike on the right. Scan the area for Least Grebe, Least Bittern, Little and Snowy Egrets, Little Blue Heron, Striated and Green Herons, Purple Gallinule, Common Moorhen, Black-necked Stilt, Wattled Jacana, Lesser Yellowlegs, Ruddy Turnstone, various species of sandpipers, Laughing Gull, Least, Yellow-billed, and Large-billed Terns, Black Skimmer, Ringed Kingfisher, Yellow-chinned Spinetail, Pied Water-Tyrant,

White-headed Marsh-Tyrant, Gray-breasted Martin, White-winged Swallow, and Yellow-hooded Blackbird. Watch for secretive species such as Masked Duck. It is often possible to observe floating nests of Least Grebe on the open ponds and it is not unusual to see Spectacled Caimans on the periphery of the ponds.

Examine the Caribbean pines on the right in which a Cape May Warbler spent several months. After departing through the gate, 'pish' near the vegetation on the right for Masked Yellowthroat.

Chaguaramas Area
10 km – Half a day

The Chaguaramas area is located northwest of Port-of-Spain near the end of the prong of Trinidad that extends towards the Boca Islands and the Paria Peninsula of Venezuela. Until recently the area was relatively inaccessible to birders because of its active military status. Since the early 1990s the Chaguaramas Development Authority has granted birders limited access to certain parts of the area. Some areas are covered with tall grasses that support populations of Sooty Grassquit.

The staff of the Trinidad and Tobago National Park Department has compiled a checklist of the birds of Chaguaramas that is updated with the assistance of visiting birders.

Permission is needed to enter the Chaguaramas area. Tour guides must pay a license fee of TT$100 to the Chaguaramas Development Authority for a license to conduct birding tours. Tour leaders must limit group size to 20 persons. Visitors are allowed to bird in and around the Chaguaramas Peninsula without a fee with advance notification to the Chaguaramas Development Authority. However, birding in the Morne Catherine area costs TT$20 per person.

Chaguaramas Development Authority
P.O. Box 3162
Carenage, Trinidad
(868) 634-4349/4227/4312/4364 voice
(868) 634-4311 fax

Stop at the Police Post upon entry because separate permission (additional to the permit) is needed to visit Tetron, Staubles Bay, the Omega Tracking Station, and the Rifle Range, all of which are restricted areas. It is possible to arrange permits by telephone before you visit.

To reach the Chaguaramas area follow the Western Main Road west from Port-of-Spain, bearing left at every turn until reaching a marina called 'Power Boats' on the left. Stop here to scan the power lines and boat masts for American Kestrel. This is the only reliable place in Trinidad and Tobago to find this species.

A guard station in the middle of the road marks the entrance to the Chaguaramas restricted area. Stop to obtain permits. Most of the roads that lead north, towards the heights of Mount Catherine, are worth birding. The coastal scrub at the western tip of the peninsula has the greatest potential in Trinidad for hosting Venezuelan strays such as Streaked Saltator and Northern Scrub-Flycatcher.

After birding the Chaguaramas area, visit Tucker Valley, another promising area in western Trinidad. This pleasant gap between mountains lies about two-thirds of the way between Port-of-Spain and Chaguaramas. On exiting the Chaguaramas area, take the first major left (Tucker Valley Road) at a Y-junction. There are many places to pull off the road and bird.

At the end of Tucker Valley Road are the remains of a World War II dance hall at Corozal Point, overlooking Maqueripe Bay. Several rugged trails lead west, up and into the scrub, where it is often possible to see stray Venezuelan species. The shiny, slippery, silvery surface is a metamorphic rock called micaceous schist.

To return to Port-of-Spain, retrace your route through the Tucker Valley, then turn left onto the Western Main Road.

BIRDING IN TOBAGO

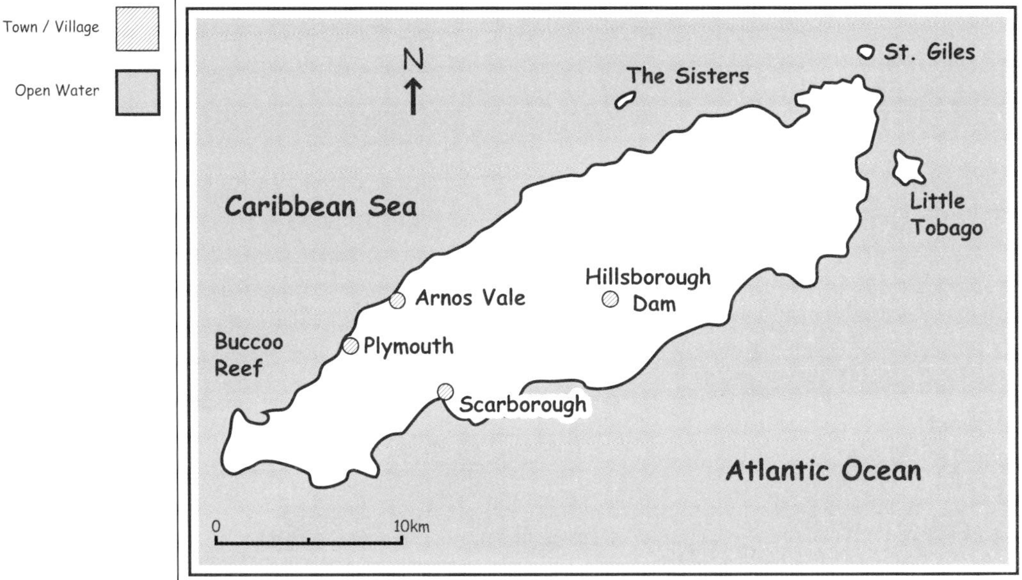

Birding in Tobago is considerably different from birding in Trinidad. Tobago is small enough to cover during a short visit. As landowners continue to abandon cultivated areas, extensive tracts of former plantations are reverting to a wild state, providing new habitat for wildlife. Unfortunately, some species, such as boobies, are still hunted despite protection under law. The good news is that there are no poisonous snakes or chiggers in Tobago.

About 90% of the human population of Tobago lives in the southern third of the island. North of the capital city of Scarborough the land becomes mountainous. Extremes of climate and habitat can be found on Tobago, with lush rainforests along the central spine and xeric-scrub deserts on the windward sides of islets such as St. Giles and Little Tobago. Although fewer species of land bird occur on Tobago than on Trinidad, with fewer individuals of each species, that aspect is offset by their greater tameness.

Imitations of Ferruginous Pygmy-Owl work poorly on Tobago because the owl does not occur there. Local guides, such as Adolphus James, can imitate the calls of nearly all species of birds in Tobago and attract them into view in that manner.

If time is limited visit the mangrove swamps in the south, the central highlands, Hillsborough Dam, and Little Tobago Island. The 13 species found in Tobago but not in Trinidad are Red-billed Tropicbird, Rufous-vented Chachalaca, Striped Owl, White-tailed Sabrewing, Red-crowned Woodpecker, Olivaceous Woodcreeper, White-fringed Antwren, Venezuelan Flycatcher, Blue-backed Manakin, Caribbean Martin (accidental in Trinidad), Scrub Greenlet, Black-faced Grassquit, and Variable Seedeater. Distances given under the Tobago site headings start at the airport.

Crown Point Airport Area
5 km – 3 hours

Crown Point Airport in Tobago provides surprisingly rewarding birding. The following species have been seen in less than an hour within sight of the runway: Brown Booby, Brown Pelican, Magnificent Frigatebird, Cattle Egret, Yellow-crowned Night-Heron, Laughing Gull, Royal, Sandwich, Roseate, Common, Bridled, and Sooty Terns, Brown Noddy, Eared and White-tipped Doves, Ruddy Ground-Dove, Smooth-billed Ani, Ruby-topaz Hummingbird, Red-crowned Woodpecker, Yellow-bellied Elaenia, Brown-crested Flycatcher, Tropical and Gray Kingbirds, Caribbean Martin, Bare-eyed Thrush, Tropical Mockingbird, Scrub Greenlet, Bananaquit, Blue-gray Tanager, Black-faced Grassquit, Carib Grackle, and Crested Oropendola.

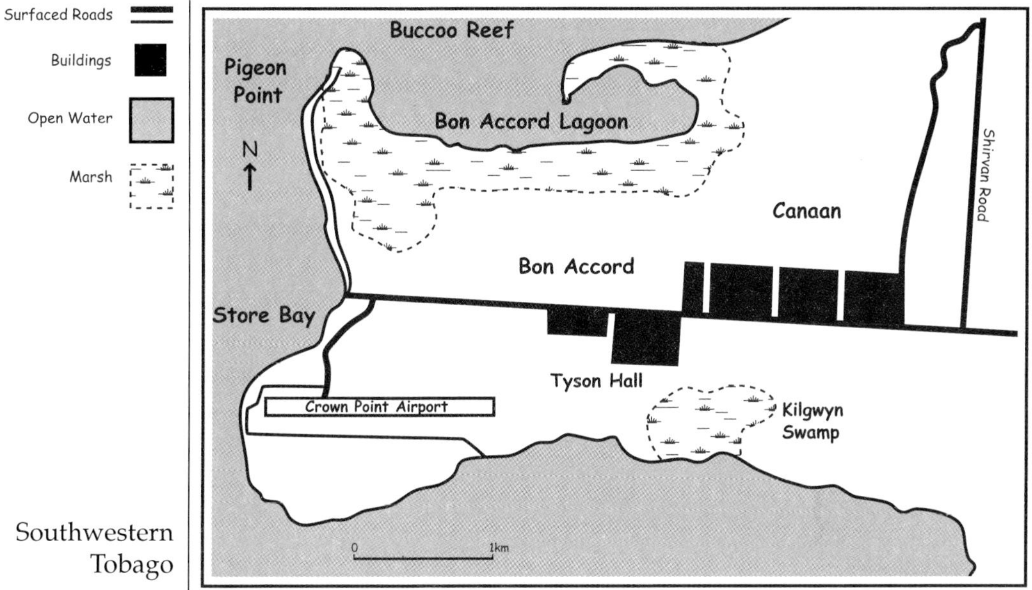

Southwestern Tobago

Drive from the airport's main exit until reaching the street perpendicular to the airport terminal. Make the first left turn, on to Milford Road, then follow the signs for Fort Milford. A road follows the perimeter of the airport property. Bird first along the beach, which is called Sandy Point, then through an open savannah to the sea cliffs on the south side of the island. Park along the beach at every access point and scan the surf for Brown Booby, Brown Pelican, Laughing Gull, and terns.

Follow the perimeter road for 2.5 km, scanning the chain-link fence for birds. Gray Kingbird, grassquits, and Ruddy-breasted Seedeater are more common when the grass grows tall. At the end of the road, walk along the conspicuous trails that lead to the sea cliffs. Scan the channel (Galleons Passage) between Trinidad and Tobago. Wherever possible, push through the thornless beach plum trees around the cattle fields to the cliffs. Continue east along the shoreline, watching for Barn Owl and White-tailed Nightjar roosting in the dense woods. It is also a good idea to explore the caves in the sea cliffs below.

Despite evidence to the contrary, legend has it that this stretch of beach is the site of Robinson Crusoe's cave.

Pigeon Point
10 km – Half a day

The Bon Accord Lagoon area is under development so this section soon may be obsolete.

Coconut Palms

A northward-pointing peninsula northeast of Crown Point Airport consists of stunningly beautiful white beaches set against a background of coconut palms. Just landward, cabanas give way to a mangrove swamp that rings Bon Accord Lagoon. The wetlands that encircle Bon Accord Lagoon provide a rare habitat in Tobago. Species found here include Pied-billed Grebe, Osprey, Rufous-vented Chachalaca, Black-bellied (Grey) Plover, American Golden-Plover, migrant North American sandpipers, Eared and White-tipped Doves, Mangrove Cuckoo, White-tailed Nightjar, Ruby-topaz Hummingbird, Copper-rumped Hummingbird, Barred Antshrike, White-fringed Antwren, Brown-crested Flycatcher, Gray Kingbird, Caribbean Martin, Bare-eyed Thrush, Scrub Greenlet, migrant North American warblers, Black-faced Grassquit, and surprisingly musical Shiny Cowbirds.

To reach Pigeon Point, turn right at the airport's main exit onto Milford Road, passing a red-and-white radio tower on the left. At 1.0 km follow the arrow on the blue-and-white sign to Pigeon Point. Drive straight ahead towards the Caribbean. This slight indentation in the western coast is called Store Bay. Follow the road around a curve to the right. At 2.1 km turn right at the sign 'Fisherman's Row' and park.

Search the edges of the mangrove-lined lake ahead for Common Moorhen and Yellow-crowned Night-Heron. Drive through the entrance gate. The entrance fee is TT$12 per person. Proceed along Coconut Drive, passing a checkpoint at 2.9 km, where a guard will tear the corner off your entrance ticket. Continue past the beachside cabanas, following the road to the right. The bright blue-green bay ahead is Nylon Pool, famous for its sandy, stoneless bottom. Its name comes from the fact that the colour of the water resembles prototypic nylon.

Bear right at the Y-junction at 3.4 km. The road degenerates here and becomes rutted and primitive but still driveable. At 3.6 km turn right at an obscure Y; a left turn here goes to the beach instead of into mangroves. If the lane is muddy, park at 3.9 km and walk into the coconut/mangrove swamp; otherwise drive on a bit farther and park. The lane winds through open mangrove woodlands for the next 0.5 km. Red mangroves dominate the swamp, with white mangroves on the southern edge. Abundant here are Yellow-crowned Night-Heron, Copper-rumped Hummingbird, Ruby-topaz Hummingbird, Brown-crested Flycatcher, Gray Kingbird, Barred Antshrike, and Shiny Cowbird.

Mangrove scrub at Pigeon Point

Try pishing and whistling the call of the Ferruginous Pygmy-Owl to attract any migrants from Trinidad or South America that may be on the island. It may also be worthwhile whistling the call of the Eastern Screech-Owl as Neotropic migrants such as American Redstart, Prothonotary Warbler, and Northern Waterthrush sometimes flash into view in response.

Throughout the swamp Rufous-vented Chachalacas and Common Moorhens create a ruckus in the mangroves, though they keep out of sight. Where the lane enters a wide, grassy pasture, cross a conspicuous ditch on the right and walk to the tip of the short peninsula that overlooks a tidal lake. Look for Little Blue, Tricolored, and Green Herons (not Striated as in Trinidad), White-cheeked Pintail, Black-bellied (Grey) Plover, American Golden-Plover, and Whimbrel. Some of the Whimbrel belong to the European race *phaeopus*, with white rump and underwings. Search carefully on the mangrove roots for sleeping White-tailed Nightjars. Large Iguanas can sometimes be seen in the crowns of the mangroves; when disturbed they hurl themselves into the water below with a resounding splash.

Back on the lane, walk to a concrete footbridge. Turn right, looking along the wet channels for shorebirds and waders. Explore all passable dikes around this wet area, noting the birds flying over, which include the ever-present Magnificent Frigatebirds and Cattle Egrets.

Pigeon Point Lagoon

Continue on, bearing left wherever possible, staying in the swamp as it turns north then west on the Golden Grove peninsula, eventually reaching the shore of the lagoon and walking back along the beach to the car.

Bon Accord Treatment Ponds 2 km – Half an hour

This site takes only minutes to bird and can be combined easily with any of the other Crown Point sites.

From the intersection of the airport/Pigeon Point roads, proceed for a mere 0.2 km, turning left onto Alfred Crescent. Bear to the right, then proceed 0.5 km. Turn right in 0.3 km onto Anthony-Charles Crescent and watch for the treatment ponds on the left. The ponds are inside a chain-link fence. Enter the pond area through the metal gate if it is open. Otherwise bird from the side of the road. Species to look for include Least Grebe, Anhinga, a variety of heron species, White-cheeked Pintail, wintering Blue-winged Teal, and shorebirds. Across the road is an area of rushes that host blackbirds and possibly rails and crakes.

After birding the pond area continue to the end of Anthony-Charles Crescent and turn left onto Roberts Street. Proceed until the road turns 90° to the right. Park on the roadside and walk straight ahead along the drainage canal towards Bon Accord Lagoon.

Along the ditch watch for herons and migrant sandpipers. The red mangroves on the right may hold Mangrove Cuckoo, White-tailed Nightjar, Brown-crested Flycatcher, Yellow Warbler, American Redstart,

Prothonotary Warbler, and Northern Waterthrush, while the brush on the left may be productive for other Tobago specialities such as White-fringed Antwren and Scrub Greenlet. At the end of the trail scan the broad expanse of Bon Accord Lagoon for terns and gulls and the shoreline for sandpipers and plovers.

To return to Milford Road, turn around and follow Roberts Street, which ends at Milford Road.

There is an excellent restaurant nearby; Singh's Roti Shop on the right just before a street called 'Friendship Ext.' offers excellent beef, goat, chicken, and vegetarian rotis, along with ice-cold drinks.

From Singh's, backtrack and turn right onto Golden Grove Road on the right beyond 'Friendship Ext.'. After 1.2 km many footpaths and old trails lead into the littoral forest on the left. Follow any of them along the shaded watercourses, heading west towards the lagoon. The same species may be found here as along the Roberts Street ditch. The area teems with White-tipped Doves. From here to its end at Shirvan Road, Golden Grove Road is rough. On reaching Shirvan Road turn left and proceed to the gated entrance to Buccoo Marsh.

Buccoo Marsh
10 km – 2 hours

Wetlands are scarce in Tobago so migrant shorebirds have few choices of places to feed. Buccoo Marsh has produced most of Tobago's records of Old World species as well as Trinidad and Tobago's first record of Wilson's Phalarope. Other species have included Least Grebe, Pied-billed Grebe, Anhinga, Scarlet Ibis on rare occasions, Fulvous and Black-bellied Whistling-Ducks, Masked Duck, Osprey, Merlin, Rufous-vented Chachalaca, Black-bellied (Grey) Plover, American Golden-Plover, a wide range of resident and migrant herons and sandpipers including many Eurasian strays, Eared and White-tipped Doves, White-tailed Nightjar, Ruby-topaz and Copper-rumped Hummingbirds, Barred Antshrike, White-fringed Antwren, Brown-crested Flycatcher, Gray Kingbird, Caribbean Martin, Bare-eyed Thrush, Scrub Greenlet, a few migrant North American warblers (including American Redstart and Prothonotary Warbler), Black-faced Grassquit, and Shiny Cowbird. It is best to visit Buccoo Marsh early in the morning when the birds are most active and the sun is behind you.

To reach Buccoo Marsh turn right at the airport's main exit onto Milford Road, passing a red-and-white radio tower on the left. Continue for about 2.0 km as the road gradually bears eastward, passing through two villages, Bon Accord and Canaan, before coming to Shirvan Road on the left. Turn left onto Shirvan Road and proceed for just over 2.0 km. After passing the junction with Buccoo Road on the right and then Golden Grove Road on the left, the road curves to the left. As it begins to straighten out and then to curve to the right, watch for a gated wire fence on the left. That fence encloses a grazed cattle field across which is Buccoo Marsh.

Park on the left, well off the road. The groundsman will take the TT$1 per person entrance fee. If no one is about, send the fee to Adolphus James (see page 160) requesting him to forward the money.

Once inside the fence, search the guava trees on both sides of the path for Barred Antshrike, White-fringed Antwren, Brown-crested Flycatcher, and Scrub Greenlet. Red-crowned Woodpeckers often are noisy and conspicuous on the trunks of the hole-ridden coconut palms throughout the area.

Most of the paths lead to the marsh but the best initial views are to the left of the main cow path. Walk through the open brush for a few hundred metres until the semicircular, grass-bordered, mangrove-backed lagoon can be seen ahead.

Along the marshy edge look for long-legged waders including Little Egret, Tricolored and Green Herons, Southern Lapwing, Wattled Jacana, and shorebirds, including both species of yellowlegs. Besides yellowlegs, the most common small shorebirds are Solitary, Spotted, Semipalmated, Western, and Pectoral Sandpipers. It is not unusual to spot a rarity here; Wilson's Phalarope has been seen feeding at the feet of a Little Egret.

In the emergent vegetation, especially along the shore on the right, look for both species of night-heron, Masked Duck, any rails or crakes that may be present, Common Moorhen, both species of coot, and Whimbrel (including the Eurasian race *phaeopus* with white rump and underwings).

In the open water scan for whistling-ducks, White-cheeked Pintail, and migrant North American ducks such as Blue-winged Teal. The mangroves that grow above the open water usually contribute at least one Anhinga, often with outstretched wings. Gray Kingbird may be found in the mangroves along with a few shorebirds resting on the mangrove roots.

Retrace your path to the opposite side of the lagoon. Working very slowly, walk around the perimeter of the lagoon, following it as it continues left. Prothonotary Warbler and Northern Waterthrush are sometimes found here. Beyond, on the small mangrove islands, Cattle Egrets may be nesting with sporadic visitors such as Snowy or Little Egret. Coming out into the open, scan the wet grass ahead and to the left for Whimbrel and other shorebirds.

Most bird tours turn back at this point because the ground gets mucky and the woods thick, but it is possible to continue. This area is particularly good for White-tailed Nightjar. The mangrove swamp ahead is contiguous with that at Pigeon Point.

One of the most interesting areas in this part of Tobago is the open water and marshy areas around Buccoo Lagoon. To reach the lagoon, on leaving the gated entrance to Buccoo Marsh continue north for 0.8 km to a coconut-tree studded golf course straddling the road. Turn left and follow the road for 0.4 km, then turn right and drive 0.5 km to Buccoo Lagoon. Check the open water for gulls and terns and the beach for Sanderlings and other shorebirds. Walk just inland from the beach towards an area of marsh vegetation, watching for Green Heron, Common Moorhen, Wattled Jacana, and other wetland species. This is the place where Tobago's first (South America's second) Western Reef-Egret was discovered. Other birds in the area may include Little Egret and the Eurasian race of Whimbrel.

Grafton Estate
25 km – Half a day

To reach the Grafton Estate continue 3.5 km past the gated entrance to Buccoo Marsh and watch on the right for a track running steeply uphill, and a sign that reads 'Grafton Caledonia Bird & Wildlife Sanctuary.' Drive to the estate house and park in the area provided in front of the Copra House on the right. Birds are fed here at 8 a.m. and 4 p.m.

Until about 25 years ago Eleanor Alefounder owned the great house on Grafton Estate. She left the windows open in every room of the house, and birds began to include the house in their flight path, flying in one side and out the other. At that time the estate was semi-abandoned, although the trails were still maintained to a modest degree and were very productive for birding. Eleanor invited visitors to tea when she fed the birds – sugar water for the Bananaquits, corn for the Rufous-vented Chachalacas, and cheese for the Blue-crowned Motmots. Nowadays birders are still welcome at Grafton Estate, although the house itself is privately owned and birders are requested not to birdwatch near the house.

Among the speciality birds of the area are Rufous-vented Chachalaca, Blue-crowned Motmot, Olivaceous Woodcreeper, White-fringed Antwren, Fuscous Flycatcher, Blue-backed Manakin, and Scrub Greenlet.

Three paths diverge from the Copra House, one to the left, one up the hill, and one to the right; all are excellent for birding. Mosquitoes can be a nuisance on the two trails that lead down into the forest.

Arnos Vale
25 km – Half a day

The Arnos Vale Hotel is located on the Caribbean side of Tobago nestled in a private bowl-shaped valley that opens to a secluded cove. The local birds have been fed for so many years that they have little fear of humans. Paths on the property and through the nearby grassy hills provide easy access to good birding areas in a variety of habitats. Visitors can also enjoy world-class cuisine.

To reach the Arnos Vale area, drive north 5.0 km from the airport, following the directions for Buccoo Marsh and continuing on Shirvan Road past Grafton Estate, which will be considered the starting point. After 1.5 km the stream on the left is good for kingfishers and herons, including Black-crowned Night-Heron, a species hard to find farther north on Tobago.

After 2.3 km turn left onto a drive just beyond the entrance to the Turtle Beach Hotel. At the end of this drive is an open building from which it is possible to scan Turtle Bay while standing in the shade. This spot is Tobago's premier site for observing gulls and terns.

From the Turtle Beach Hotel continue 0.8 km and take the left turn towards Plymouth. After 0.8 km, stay to the right at the fork onto Arnos Vale Road. The substantial waterway that drains the area, and over which the bridge crosses, is the Grand Courland River.

Arnos Vale is 1.7 km farther, on the left. Park near the white-gated entrance on the left. Bird the lane straight ahead or turn left and enter the hotel grounds. Leave your vehicle in the parking area to the left of the guard station. The Arnos Vale staff welcome birders, especially for the bird feeding at 4 p.m. Birds that frequent the grounds and bird feeders include Rufous-vented Chachalaca, Ruby-topaz Hummingbird, Red-crowned Woodpecker, a variety of tanagers, and the two resident grassquits (Blue-black and Black-faced). A Belted Kingfisher often spends the winter on the rock face to the right of the lagoon, and Brown Boobies are usually visible farther out.

Another interesting site in the Arnos Vale area is the Arnos Vale Waterfall Restaurant and Nature Park. Turn left on leaving the Arnos Vale Hotel and look for the entrance on the right after 1.8 km. The restaurant was built around a waterwheel and partially excavated sugar mill. It is located adjacent to the Arnos Vale River amid hundreds of heliconia and wild gingers shaded by towering bamboos and flowering trees.

Belmont Road
68 km – 1 day

Belmont Road is close to bustling Scarborough but is little travelled and wide, with lots of places where it is safe to stop. The road begins next to the Atlantic and winds uphill through open, grassy areas with forests nearby. The diverse habitat and proximity to Scarborough make this road attractive to birders.

To reach Belmont Road, from Crown Point Airport drive to Scarborough via the four-lane Claude Noel Highway. Continue north about 6.5 km beyond Scarborough. After driving past most of Hillsborough Bay beach, turn left onto Belmont Road just before the concrete bridge. The best birding is along the first 4 km. Farther on the land is more densely populated and the surroundings more open.

Be sure to bear left at 3.0 km to stay on Belmont Road. Watch for raptors.

Hillsborough Dam and Old Castara Road 64 km – 1 day

Hillsborough Reservoir

Before visiting this site, it is necessary to obtain a WASA permit. See the section on the Arena Forest (page 53) for instructions.

This tour explores the mountains of the Main Ridge, first Hillsborough Dam, a freshwater reservoir, then deeper into the forest along an ancient rustic trail. On this trip it is possible to see Least Grebe, Blue-crowned Motmot, Rufous-tailed Jacamar, Blue-backed Manakin, and about 40 other species of birds.

From Crown Point Airport turn right and drive to Scarborough via the four-lane Claude Noel Highway, following the signs to Scarborough.

At 10.4 km turn left and drive 0.3 km to the summit of the hill for a scenic view of the capital city of Scarborough, which hugs Rockly Bay. Return to the Claude Noel Highway, turn left, and continue past Scarborough for several kilometres as the expressway narrows to a two-lane road. On entering the village of Mount St. George, cross a bridge with the Atlantic Ocean on the right. Watch carefully on the left for a large wooden store followed immediately by a white house at a Y-junction. Turn left onto this steep unmarked paved road (Mount St. George-Castara Road) leading to Hillsborough Dam. Another landmark to watch for as you start to climb the hill is a sign on the right that reads 'Notice, Tobago House of Assembly'. If you miss this turn, take the next left, as that road intersects the Mount St. George-Castara Road a few kilometres farther.

The pavement ends beyond Mount St. George. The actual mount is on the left as you climb this steep road. Pass a quarry and stay to the right at the Y-junction. Watch for Orange-winged Parrot, Blue-crowned Motmot, and Rufous-tailed Jacamar flying across the road and for Short-tailed and Gray-rumped Swifts plummeting along the cliff faces. The road eventually comes to a group of beige-and-blue buildings – the water control station for Hillsborough Dam. Park here and check the powerlines along the road for Gray Kingbird and Caribbean Martin, the Croton bushes for nesting Black-throated Mango, and the reservoir for Least Grebe, Neotropic Cormorant, Anhinga, and Gray-rumped Swift. At the dam itself watch for migrant swallows including Bank Swallow (Sand Martin) and Barn Swallow. Spectacled Caimans can be seen basking along the margin of the reservoir.

Continue driving up the hill beyond the dam. Stop wherever paths lead into the woods to explore the grassy fringes for seedeaters and grassquits, the forest edges for Collared Trogon, Red-rumped and Golden-olive Woodpeckers, Stripe-breasted Spinetail, Yellow-bellied Elaenia, Fuscous Flycatcher, and White-winged Becard; and the deeper rainforest for Olivaceous Woodcreeper, Gray-throated Leaftosser (rare), White-fringed Antwren, and Blue-backed Manakin.

At the next junction turn right onto Old Castara Road, park, and bird on foot. This road is one of the best high altitude sites for birding in Tobago except when there is too much wind, when the bamboo becomes noisy. Pishing often attracts flocks of up to a dozen different species.

Louis D'Or River and Beach
1 km – 1 hour

To reach the Louis D'Or River from Crown Point, drive past Scarborough and watch for mile marker 18½ along the road (a black-- and-white concrete post showing 18 and two black bars) and park off the road. Walk slowly onto the bridge and scan for the resident Green Kingfishers. During migration it is possible to find sandpipers, and herons are likely in any season.

Look into the water from the bridge. The streamlined, rather attractive fish that first catch your eye are mullet. Continue gazing at the bottom and eventually a fairly large tadpole-like fish will move. This is a goby.

Return to your car and drive down the road towards the beach. Grassquits are conspicuous in the taller grasses and Palm Tanagers flit among the palm fronds. Drive to the end of the road at the beach and park. Scan the beach in both directions. A short walk towards the mouth of the river may provide views of Little Blue and Tricolored Herons, Yellow-crowned Night-Heron, several species of shorebirds, and migrant North American warblers.

Little Tobago (Bird of Paradise) Island 95 km – 1 day

Windward side of Little Tobago Island

It is best to obtain prior permission to visit Little Tobago Island from the Forestry section of the Department of Agriculture, Forestry and Fisheries of the Tobago House of Assembly. However, if you hire a boatman in Speyside to take you there he will almost certainly have the required permit. In any case, information can be obtained from the Botanic Station, Scarborough (838-639-3265) or from the Forestry Office, Windward Road, Mount St. George, Tobago.

Little Tobago Island, one of Trinidad and Tobago's most impressive wildlife sanctuaries, lies about 2 km off the northeastern coast of Tobago. The island comprises only 13 ha (280 acres), much of it very hilly with a maximum elevation 140 m (460 ft.). Fan palms and gumbo-limbo trees predominate in the forest. Cactus and other xeric shrubs dominate the cliff-sided eastern (windward) side. The island's main attraction for birders is its essentially pelagic location. It is impossible elsewhere on Trinidad and Tobago to obtain such close views of nesting Red-billed Tropicbird, Brown Booby, Laughing Gull, Bridled and Sooty Terns, and Brown Noddy. Red-footed Booby and Magnificent Frigatebird nest on nearby St. Giles and can be seen in the distance from the overlooks on the northeastern side of Little Tobago Island.

The alternate name of the island, Bird-of-Paradise Island, was derived from the introduced Greater Bird-of-Paradise *(Paradisaea apoda)*. This species, formerly the focal point of birding on Little Tobago Island, was practically extirpated by Hurricane Flora in 1963. The species has not been recorded there since the early 1980s. It is rumoured that the last individual of this colonial species was observed attempting to blend in with a colony of Crested Oropendolas, which would have nothing to do with the much larger species. On a more positive note, more than 50 other species of birds occur regularly on Little Tobago Island.

The trip from southern Tobago to Little Tobago Island and back takes an entire day. To allow the maximum amount of time to be spent on the island, an overnight stay at the north end of Tobago at the Blue Waters Inn or Turpin's Cottages is recommended. Since birds are most active just after dawn and the sea at that time is usually fairly calm, those travelling from the Crown Point area will need to leave their hotel several hours before dawn. To reach Little Tobago Island from southern Tobago, follow the Claude Noel Highway north past Scarborough. Continue for about 40 km, staying on the main road. Opposite a fire station on the right, the Roxborough-Bloody Bay Road climbs to the Tobago Main Ridge Preserve, the oldest nature reserve in the Western Hemisphere. If choosing to detour for rainforest birding, take this scenic, well-paved drive to the cool, moist heights. The White-tailed Sabrewing, which was nearly extirpated by Hurricane Flora in 1963, can be found throughout this area.

Other target birds are Great Black-Hawk, Collared Trogon, Stripe-breasted Spinetail, Venezuelan Flycatcher, Blue-backed Manakin, and Yellow-legged Thrush. Among the more conspicuous butterflies is a lovely black-and-green swallowtail called the White-tailed Patch. Almost any trail that leads off the road is productive. Most birding groups proceed directly to the summit and bird the well-marked Gilpin Trace. Roxborough Valley Trace and Branch 2 Trace experience much less birding pressure and thus are highly recommended.

After hiking, drive a few kilometres west and enjoy a spectacular view of northwestern Tobago from the Bloody Bay Recreational Site forestry building, then return to Roxborough and turn left.

Licenced Tobago nature guides

Bloody Bay Recreation Site

Continue through the town of Roxborough and about 12 km farther drop down into tiny Speyside.

There are several options for making the 2-km trip across Tyrrell's Bay to Little Tobago Island. The manager at the Blue Waters Inn can arrange in advance to have a boat ready to take you there or you may opt to negotiate with the fishermen in Speyside upon arrival. In any case, if you arrange with a boatman to be dropped off and then picked up later, be sure to confirm the time he will return. Take extra provisions in case rough seas prevent the boatman from returning as arranged. Be sure to take a supply of drinking water as none is available on the island. Rates for the round trip vary considerably depending upon the number of persons travelling together. At the time of writing a typical fare was about US$20. The fare increases if the boatman waits for the return journey. Be sure to wrap your binoculars, camera, and other optical equipment in waterproof material, preferably within nested plastic zippered bags, to protect them from salt spray. While crossing watch for Hawksbill, Leatherback, and other species of sea turtles that nest along the coasts of Trinidad and Tobago from late February to late August and for large barracudas. Do not trail your hands in the water.

The brief crossing is occasionally rough, especially during changing tides, but exploring Little Tobago Island and viewing the colonies of seabirds is well worth the slight discomfort.

Once ashore do not be alarmed by the loud rustling and scratching in the forest. The source is usually one of the hundreds of feral chickens on the island. A set of concrete steps lead up the hill to the warden's station. The burrows in the soft earth in the banks are dug by Audubon's Shearwaters, which arrive and depart from their nests at night. Cautious use of a torch may provide a view of a shearwater at the far end of its burrow. Do not get too close as shearwaters bite.

Continue uphill to the warden's house. Check the visitors' log to see if anyone else is on the island. If an official is present, show the permit. If the house is vacant, complete the visitor's log. Before departing from the island, it is courteous to leave a list of any unusual sightings at the warden's house.

From the warden's house continue east, watching for White-tailed Nightjar (conspicuous at dawn and dusk only), until reaching the cactus-covered cliff on the eastern side of the island. On the cliffs to the left, are nesting colonies of Red-billed Tropicbird, Brown Booby, Laughing Gull, Sooty and Bridled Terns, and Brown Noddy. It is possible to photograph these species at close range. December to February marks the peak nesting period for Red-billed Tropicbird. Brown Boobies nest mostly in July and again in January and February. Sooty Terns nest during February and March, Brown Noddies in March, and Laughing Gulls in May. If the view seems familiar, it might be because this is where the tropicbird segment of Sir David Attenborough's *The Trials of Life* was filmed.

When not nesting on Little Tobago Island, Red-footed Boobies can usually be seen flying east over the long, thin island that lies about halfway to the horizon. Watch carefully for any brown boobies with white rumps; those are Red-footed Boobies.

Rufous-tailed Jacamar

Retrace your path past the warden's house and turn left on the conspicuous trail that leads uphill along a ridge. Species such as Broad-winged Hawk (March through May), Pale-vented Pigeon, White-tipped Dove, Red-eyed ('Chivi') Vireo, and Crested Oropendola may be seen here. Watch for Roseate Terns feeding offshore. Bridled and Sooty Terns are also common just off the island. Blue-crowned Motmots are easy to find, and Venezuelan Flycatchers are particularly abundant but sedentary. Seedeaters are rare throughout Trinidad and Tobago, but Lesson's and Yellow-bellied Seedeaters reportedly are possible to find on Little Tobago from May until October.

SELECTED BIRD LIST

The following list includes species of unusual status or patchy distribution and those frequently missed by visitors. It excludes widespread species and most accidentals while emphasizing the birding locales described in this book. As far as possible, bird names and their arrangement follow the American Ornithologists' Union's Check-list of North American Birds, 7th edition, 1998 (Am. Ornithol. Union, Washington, D.C.).

Little Tinamou
A pigeon-sized, upright, strictly terrestrial species that prefers second-growth forests with a high, dense canopy; Mount St. Benedict, Arena Forest, Brigand Hill. Responds to taped recordings of its call but not to vocal imitations. Calls throughout the day. Commonly heard but almost never seen.

Least Grebe
Trinidad: breeds at Trincity Ponds. Also reported from Caroni rice fields, Caroni Swamp, Brickfield ponds, Pointe-à-Pierre Wildfowl Trust, Pitch Lake, Hollis Reservoir. Tobago: Bon Accord and Lowlands sewage ponds, Buccoo Marsh, and Hillsborough Dam. Apparently more common now than formerly. V-shaped wake is more conspicuous than the tiny bird itself.

Pied-billed Grebe
Trincity Ponds, Caroni rice fields.

Shearwaters and petrels
Almost never seen from shore in Trinidad anywhere but at Manzanilla Beach on the east coast, but might be possible to see from shore in Tobago. Also possible to observe at rare intervals from the Trinidad-Tobago ferry.

Cory's Shearwater
Rare vagrant. Trinidad-Tobago ferry, Galera Point.

Greater Shearwater
Vagrant. Trinidad-Tobago ferry. Also seen several times in recent years in small numbers from an oil-drilling platform 24 km east of Galeota Point June 14–17; should also be looked for off Manzanilla Beach during that period.

Manx Shearwater
Manzanilla Beach (mostly dead birds).

Audubon's Shearwater
Trinidad-Tobago ferry. Breeds on Little Tobago Island. Comes and goes from its burrow at night and thus is difficult to observe.

Leach's Storm-Petrel
Trinidad-Tobago ferry. Seldom seen from shore. February to early June.

Galera Point, Balandra, and along Manzanilla Beach, where occasional individuals are found sick or dead. The most promising spot is the first pullout on the left after the road descends to the beach from Lower Manzanilla, just after the bridge. Persistent scanning usually pays off.

White-tailed Tropicbird
Rare. Little Tobago Island.

Red-billed Tropicbird
A speciality of Little Tobago Island, breeding on cliffs on the ocean side; hard to find anywhere else but off northern Tobago. Has been seen from the Trinidad-Tobago ferry and off Galera Point.

Masked Booby
Began breeding on St. Giles in the late 1990s.

Brown Booby
Very rarely seen from mainland Trinidad but common offshore in Tobago, especially in the north. Seen at rare intervals from the Trinidad-Tobago ferry.

Red-footed Booby
Very rarely seen in Trinidad and uncommon offshore in Tobago; breeds on St. Giles and Little Tobago Island. Hard to find anywhere else except far off to the north of Little Tobago Island; on dark morph look for the white diamond formed by the feathers on the rump and lower back. Brown Booby has a dark rump.

Brown Pelican
Distribution is expanding; undergoing a population explosion in Trinidad.

Neotropic Cormorant (formerly Olivaceous Cormorant)
Pointe-à-Pierre Wildfowl Trust (1,000+); singles are often seen in the Caroni Swamp, Hollis Reservoir, Arena Reservoir, and the San Fernando Public Fish Market. Apparently prefers saltwater for fishing, freshwater for roosting.

Anhinga
Trinidad: Caroni Swamp, Point-à-Pierre Wildfowl Trust, Caroni rice fields, Arena Reservoir, Nariva Swamp. Tobago: Hilton Hotel, Bon Accord sewage ponds, Buccoo Marsh, Hillsborough Dam. Prefers freshwater lakes with dense, overhanging foliage.

Magnificent Frigatebird
Breeds on St. Giles.

Pinnated Bittern
Caroni rice fields, Nariva Swamp (Kernahan Trace), Caroni Swamp, Aripo Livestock Station; uncommon, inconspicuous, remains motionless with bill pointed skyward at about a 70° angle. Most easily seen in the late afternoon.

Stripe-backed Bittern
Waterloo, Brickfield, Caroni Swamp, El Socorro Marsh; uncommon, inconspicuous, increasingly rare. Most visible in late afternoon. Straw coloured back with conspicuous dark streaks and lack of contrast between back and wings help separate it from Least Bittern. Ratio of sightings of Stripe-backed to Least is about 1 in 5.

Least Bittern
Increasingly rare; Caroni rice fields, Nariva and Caroni Swamps. Resident Caroni Swamp population consists of about 20 individuals. Most easily seen in late afternoon.

Rufescent Tiger-Heron
Rare and hard to find; the only recent reports are from Aripo Savannah east of Cumuto, Nariva Swamp, and south Trinidad. Try Nariva or Caroni Swamps at twilight, especially the marshes at Caroni Swamp near the boat docks. Prefers lowland marshes and swamp/mangrove forest; hides in tall grasses and sedges like a bittern.

Grey Heron
Vagrant from Europe. Trincity Ponds, Caroni rice fields. Recorded in 1959 and again in 2000–2001. Easily confused with Great Blue Heron but much shorter. White (not rufous) thighs and conspicuous white carpal patch confirm identification of this Eurasian species.

Cocoi Heron
Aripo Livestock Station, Caroni rice fields, Caroni Swamp, Arena Reservoir, Pointe-à-Pierre Wild Fowl Trust, Nariva Swamp; watch for it to fly across open channels. Prefers margins of salt marshes, freshwater swamps, and lakes; solitary.

Little Egret
Trincity Ponds, Caroni rice fields, occasionally in the few wetlands in Tobago, especially Buccoo Marsh and Buccoo Lagoon. Increasing in numbers in the southeastern Caribbean, with scores of records for the New World; many records from Trinidad and Tobago. As many as five at once in Tobago June 1986. Like Snowy Egret but with two long plumes (seasonal), blue or grey lores, duller yellow feet. Some taxonomists consider this species conspecific with Western Reef-Egret.

Western Reef-Egret
First South American record (dark morph) was in Nariva Swamp, Trinidad, in January 1986; second was in Buccoo Marsh, Tobago, December 2000 until December 2001. Similar to Little Blue Heron but with white chin, black legs, yellow feet. Some taxonomists consider this species conspecific with Little Egret.

Reddish Egret
Unknown to local guides but seen twice in recent years, mid-January and October.

Green Heron
Trinidad: Trincity Ponds, Caroni Swamp. Tobago: Buccoo Marsh. Striated Heron was lumped briefly with Green Heron by taxonomists but is now regarded as a separate species once again. Adults of the two species are easily distinguished. Striated Heron is light grey on the face, neck, and breast whereas these areas are reddish-brown on Green Heron. Striated is overall much lighter in colour than Green Heron and has a well-defined black cap. Immatures may be impossible to separate in the field. Both species occupy the same wetland habitat but Striated Heron is almost unknown in Tobago.

Striated Heron
See Green Heron.

Boat-billed Heron
Cacandee, Caroni rice fields, Caroni Swamp. Roosts in mangroves are inaccessible and the species is nocturnal.

Scarlet Ibis
Guaranteed on Caroni Swamp boat trips at dusk. Also Caroni Swamp, Nariva Swamp, South Oropuche Swamp. Also possible as fly-overs at El Socorro Marsh. Numbers fluctuate from year to year.

Glossy Ibis
Recent population expansion, now regular (usually single birds) at Caroni rice fields, near San Fernando, Caroni Swamp (southern), and along the southwest coast. Easily confused with Green Ibis.

Roseate Spoonbill
Rare; Pointe-à-Pierre Wildfowl Trust, Caroni Swamp; flocks of 10 or more possible on Caroni Swamp boat trips but irregular. Very conspicuous and hard to miss if present. Roosts in trees and feeds on mudflats or in shallow water.

Lesser Yellow-headed Vulture
Nariva; possible recent sighting of a lone bird, identification based on yellow head and low, level flight. Most (all?) other reports have been misidentifications of resident yellow-naped (Venezuelan) race of Turkey Vulture. Lesser Yellow-headed Vulture has white shafts to the upper surface of the primaries but is otherwise extremely similar in appearance to a Turkey Vulture. Unlike Turkey Vulture, very partial to marshes and wet savannahs.

King Vulture
Rare. Northern Range. Recent records of adults and immatures near Blanchisseuse, Heights of Aripo, village of Brasso Seco in central Northern Range.

Greater Flamingo
Mudflats along Gulf of Paria between San Fernando and Brickfield.

Horned Screamer
Extirpated; resident in Nariva Swamp until 1910, last recorded in 1964.

White-faced Whistling-Duck
Caroni rice fields. Formerly extirpated through overhunting; breeding stock at Pointe-à-Pierre Wildfowl Trust may account for recent sightings.

Black-bellied Whistling-Duck
Trinidad: Caroni rice fields, Caroni and Nariva Swamps. Tobago: Buccoo Marsh. Pointe-à-Pierre release programme re-established breeding population. Frequents open water, marshes with open pools, large bodies of water with grassy edges and islands; also rice fields.

Fulvous Whistling-Duck
Trinidad: Caroni rice fields, South Oropuche Lagoon. Tobago: Buccoo Marsh. Increasing in Trinidad as a result of release programmes; still rare. Prefers marshland with open pools and high grasses or lakes with grassy edges and islands; usually found in mixed flocks with Black-bellied Whistling-Duck.

Comb Duck
Caroni rice fields; accidental but when found may appear in small flocks.

Muscovy Duck
Occasionally reported from Nariva and southernmost swamps (Icacos); widespread domestication precludes determination of origin of such individuals.

Blue-winged Teal
Trinidad: Caroni Swamp, Caroni rice fields. Tobago: Buccoo Marsh.

White-cheeked Pintail
Rare. Trinidad: Caroni rice fields, Point Lisas. Tobago: Buccoo Marsh, Pigeon Point, Kilgwyn Swamp. Fully protected but very wary. Breeds at Pointe-à-Pierre Wildfowl Trust.

Masked Duck
Uncommon. Trinidad: El Socorro Marsh, Trincity Ponds. Tobago: Buccoo Marsh. Rarely seen; sits very low in the water and submerges rather than taking flight. Prefers areas with dense growth of water hyacinth and other herbaceous plants.

Gray-headed Kite
Increasing, common but local; most often seen in Northern Range, especially lower Arima Valley.

Hook-billed Kite
Southern Trinidad: Penal, Debe, Guayaguayare. Rare in wet savannahs and swamps; possibly overlooked because of sedentary nature. Recent record of three individuals at Waller Field.

Swallow-tailed Kite
Breeds in northeastern Trinidad: Cumaca, Sans Souci, Montevideo.

Pearl Kite
Uncommon. Increasingly urban: Pointe-à-Pierre, Chaguanas, Diego Martin, campus of University of the West Indies, eastern edge of Arima, Waller Field, along the Churchill-Roosevelt Highway east of Orange Grove Road, along Mayaro Road, and has bred at Pointe-à-Pierre Wildfowl Trust. Partial to coconut palms and telephone lines. Behaviour similar to that of American Kestrel.

White-tailed Kite
Formerly rare but increasing in Nariva (and possibly Caroni) Swamps; possibly Waterloo. Characteristic hovering is distinctive at a great distance.

Snail Kite
Caroni rice fields.

Double-toothed Kite
Most frequently reported from the Northern Range. Uncommon but frequent in Arima Valley; small, fast-flying hawk whose flight closely resembles that of a Sharp-shinned Hawk.

Long-winged Harrier
Uncommon. Caroni rice fields (especially across the highway from the Caroni boat docks), Cacandee, and South Oropuche Lagoon. Probably overlooked because of similarity to low-flying Turkey Vulture. Behaviour similar to that of other harriers – flies with bounding low flight, hovers, then drops on prey. Persistent scanning usually turns up at least one individual.

White Hawk
Possible to spot from most overlooks in Northern Range; less common to the south. Name often puzzles birders because it is not all-white; wings and most of tail are black.

Common Black-Hawk
Widespread and abundant, particularly in Northern Range. Often confused with Great Black-Hawk. As ffrench aptly points out, in flight best separated from Great Black-Hawk by its call, a series of 'spinking' notes, quite high-pitched for a large bird. Pairs of Common Black-Hawks often circle together; because females are noticeably

larger than males, the male often goes down on checklists as Common Black-Hawk and female as 'obvious' Great Black-Hawk. Study tail pattern (see following species).

Great Black-Hawk
Trinidad: Morne Bleu and other high elevations in the Northern Range, also reported over forested areas of Nariva Swamp. Tobago: Main Ridge, Speyside. See previous species. Common Black-Hawk has black tail with white band at base, whereas Great Black-Hawk often appears to have white tail with black terminal band. Call is a thin scream, reminiscent of a Red-tailed Hawk.

Savanna Hawk
Aripo Livestock Station, Waller Field, Aripo Savannah, Mayaro Road and throughout central Trinidad. More common at low elevations than higher. ffrench considered it rare but now more accurately described as locally common.

Rufous Crab-Hawk
Rare. Along the Gulf of Paria: Carli Bay, Caroni Swamp. A few records from Nariva Swamp and one from Pointe-à-Pierre. Similar in appearance to an immature Savanna Hawk. Sedentary.

Black-collared Hawk
Rare. Seen mostly along Mayaro Road (milepost 46 at boat trail to Bush Bush) and in Nariva Swamp. Wooded swamps.

Broad-winged Hawk
Uncommon but increasing in Trinidad; the most frequently seen raptor in Tobago.

Short-tailed Hawk
Southwestern Trinidad: Orange Valley, Point-à-Pierre, Fullerton ('Fullarton' on maps) Swamp, San Fernando. Overlooked although regular and widely distributed; often soars with vultures. Dark facial pattern resembles that of Peregrine Falcon.

Swainson's Hawk
Vagrant. Tobago: Speyside, possibly Little Tobago Island.

White-tailed Hawk
Overlooked; distribution same as that of Short-tailed Hawk but more of a lowland species, less common in Northern Range. Hovers. As with Short-tailed Hawk, dark facial pattern resembles that of Peregrine Falcon.

Zone-tailed Hawk
Overlooked or much more common than formerly, widespread; Heights of Aripo, Waller Field, Mount St. Benedict, Nariva Swamp. Closely resembles and mimics Turkey Vulture. Watch for 'vulture' stooping on

passing Turkey Vultures, from which it is easily distinguished by bright yellow cere and feet; tail bands may be obscure. Often cruises just above treetops, against which background it is difficult to discern.

Black Hawk-Eagle
Southern Trinidad. Scant records from Waller Field, Nariva Swamp, Central Range.

Ornate Hawk-Eagle
Often nests on grounds of Asa Wright Nature Centre and elsewhere in Arima Valley, where it is seen almost daily, especially before 9 a.m. (earlier than most other soaring raptors). Light underwings and narrow black tail bands distinguish it from locally common Gray-headed Kite, which has dark underwings and wide black tail bands.

Yellow-headed Caracara
Dramatic population increase in recent years; Caroni and Nariva Swamps, Waller Field, Aripo Livestock Station, Aripo Savannah. Unmistakable. No other raptor found in Trinidad or Tobago is creamy yellow with dark wings, striking pale patch in wing. Frequently pursued and harried by Tropical Kingbird.

Merlin
Uncommon but regular winter visitor to both islands; widespread over open areas (sugar cane fields, swamps) where it pursues and captures prey on the wing. Distinctive flight – about two deep wingbeats per second, almost no soaring or gliding; when in active pursuit of prey (often bats) wingbeats are so fast as to appear swallow-like.

Aplomado Falcon
Very rare; Caroni rice fields, Caroni Swamp south to South Oropuche Lagoon, Waller Field. Recent sightings on Tobago at Hilton Ponds. Increasing number of reports, especially of individuals pursuing Dickcissels in Nariva Swamp, may be a result of successful captive release programme in Texas as Aplomado Falcons follow Dickcissels on migration.

Bat Falcon
Uncommon but regularly seen from overlooks along Blanchisseuse Road, Lalaja Road, upper Cumaca Road (on left just past Valencia), and Heights of Aripo as it dashes after bats and songbirds. Most easily seen at dawn and dusk but active throughout the day. Often perches on treetops. Appears all black, Merlin-like flight.

Peregrine Falcon
Uncommon but regular visitor to both islands, widespread in open areas but particularly fond of mudflats, where it pursues shorebirds; Caroni Swamp, Aripo Livestock Station.

Trinidad Piping-Guan
Extremely rare. Morne Bleu and about 1 km south of Grande Riviere on Monte Video Road. Small populations are still found in other areas of

the remote upper reaches of Northern Range including Heights of Aripo, Cumaca, Matura: also reported from southeast (Trinity Hills). Sightings are usually of solitary birds except at Grande Riviere, where occasional flocks containing as many as six individuals can be seen.

Rails and crakes
Generally difficult to observe because of habit of skulking on ground in dense vegetation, primarily in wetlands but also in grasslands. More common than realized. Learn calls from tapes, ffrench's descriptions, or firsthand and record far more by ear than by eye. The use of playback equipment is the best method for determining identity of mystery singers.

Clapper Rail
Cacandee, Waterloo, and along tidal creeks islandwide – prefers fiddler crabs as prey. Easiest rail/crake to see because of its large size and unsecretive behaviour. Call, loud series of 3-8 grating notes – *'kak, kak, kak'* or *'chek, chek, chek'*.

Rufous-necked Wood-Rail
Cacandee, El Socorro Marsh. Very rarely seen or heard.

Gray-necked Wood-Rail
South Oropuche Lagoon, Nariva and Caroni Swamps, and breeds around reservoirs at Pointe-à-Pierre Wildfowl Trust but rarely seen. Slightly larger than Clapper Rail, not likely to be mistaken for one. More secretive in Trinidad than in other parts of its range. See ffrench for a description of its call.

Sora
Abundant winter visitor. Trinidad: Brickfield, Waterloo. Tobago: Bon Accord sewage ponds.

Ash-throated Crake
Very rarely seen or heard. As with Rufous-necked Wood-Rail, only recent records are from El Socorro Marsh.

Yellow-breasted Crake
Often the most abundant wetland crake species (except Sora in winter). Marshes south of Caroni Swamp boat docks on Uriah Butler side of trail, El Socorro Marsh. Diagnostic marks – dark eye-line below pale supercilium – must be noted quickly before bird skulks back into vegetation. Call described as frog-like, high-pitched double note accented on the first syllable, *'TEE-di'*.

Spotted Rail
Nariva Swamp; considered fairly common by local guides but difficult to observe except in early morning.

Purple Gallinule
Trincity Ponds, Caroni rice fields, and wet fields bordering Caroni,

Nariva, and South Oropuche Swamps. Uncommon, much less confiding than Common Moorhen, with which it shares habitat. Seeks cover when approached, hides in thick vegetation.

Azure Gallinule
Known primarily from a small colony along Kernahan Trace in southern Nariva Swamp, in marsh vegetation on both sides of the road just after it emerges from the coconut plantation. Also found in small numbers in the Caroni rice fields. Resembles immature Purple Gallinule but smaller. Often flies short distances. Also called Azure-winged Gallinule.

American Coot
Bon Accord sewage ponds.

Caribbean Coot
Buccoo Marsh.

Sungrebe
Elusive and little-known species, frequents heavily-forested areas around lakes, dams, streams in central Trinidad. Recent record of four birds at Hollis Dam.

Limpkin
Caroni rice fields, Caroni and Nariva Swamps, marshes in southern Trinidad. Uncommon; declining from hunting pressure. Frequents fresh water with large snails on which it feeds. More often heard than seen; ffrench's description apt – a loud, wailing scream, *'krraow'*, often heard in background of jungle films.

Southern Lapwing
Population explosion in recent years. Increasingly abundant at Aripo Livestock Station, Waller Field, Aripo Savannah, Nariva Swamp, Piarco Water Treatment Plant, South Oropuche Lagoon.

Plovers and sandpipers
Trinidad: Caroni rice fields, Caroni Swamp mudflats, Pointe-à-Pierre mudflats, Waller Field. Tobago: Buccoo Marsh, Hilton Ponds. Most North American species have been recorded; species range from abundant during migration to accidental at other times with some individuals (non breeders) remaining through summer.

Collared Plover
Uncommon and local in season. Caroni rice fields, Aripo Livestock Station, and other wet, grassy areas.

Wilson's Plover
Recent breeding record at Pointe-à-Pierre Wildfowl Trust.

Willet
Waterloo, fly-bys at Galera Point.

Upland Sandpiper
Caroni rice fields.

Buff-breasted Sandpiper
Trinidad: Caroni rice fields. Tobago: grassy eastern end of Crown Point Airport runway.

Jaegers (skuas)
Occasional at Waterloo, sometimes seen from the Trinidad-Tobago ferry.

Gulls and terns
The greatest concentrations of gulls and terns can be found in Trinidad at Waterloo and at the Trincity Ponds (terns only), Icacos Point in southwestern Trinidad; in Tobago in Turtle Bay, behind the Turtle Beach Hotel.

Black-headed Gull
Formerly regarded as accidental, now regular in small numbers at Turtle Bay, behind Turtle Beach Hotel, Tobago.

Lesser Black-backed Gull
Recent range expansion has yielded many individuals at Waterloo.

Sandwich Tern
Uncommon along coasts of both islands; regularly misidentified as Cayenne Tern because bill of this race is entirely yellowish-orange, unlike yellow-tipped black bill of northern form. Both types breed in Netherlands Antilles (morphs of same species?). Watch for it resting on fishing boats behind Turtle Beach Hotel, Tobago.

Roseate Tern
Trinidad: Galera Point. Tobago: Plymouth (nesting colonies just offshore), Batteaux Bay (Speyside area). Uncommon but regular along northeastern Trinidad and all coasts of Tobago.

Yellow-billed Tern
Trincity Ponds, Caroni rice fields, Piarco Water Treatment Plant, Arena Reservoir. According to Dr. Floyd Hayes of the University of the West Indies: 'Contrary to statements in several popular field guides, the Yellow-billed Tern has an entirely yellow bill only in alternate plumage; immatures and basic plumaged adults possess a dark tip to the bill plus an extensive dark area around the nostrils and upper ridge of the bill. Least Terns have a dark tip to the bill in alternate plumage and an entirely or mostly dark bill in immature and basic plumages; to our knowledge, no alternate plumaged Least Terns have been observed in Trinidad and Tobago. Yellow-billed Terns are also slightly larger with a proportionately larger bill than Least Terns.'

Bridled Tern
Trinidad: Galera Point. Tobago: Common at times along coast, breeds on Little Tobago Island.

Sooty Tern
Trinidad: Galera Point. Abundant at times along the coast of Tobago, breeds on Little Tobago Island.

Large-billed Tern
Same areas as Yellow-billed Tern; most common at Trincity Ponds and Caroni Swamp.

Brown Noddy
Abundant at times but only off Tobago; very few sightings from Trinidad. Breeds on Little Tobago Island.

Black Skimmer
Common but local; Waterloo, Pointe-à-Pierre.

Feral (Rock) Pigeon
Not mentioned in ffrench or on many checklists but locally common around barns; not so urban as in North America and Europe.

Pale-vented Pigeon
Uncommon and local in Trinidad; most easily seen in flight in Caroni Swamp after 4:30 p.m. Size, shape, and flight style similar to Feral Pigeon. In Tobago, can be found at Grafton Estate, along road to Hillsborough Dam, and on Old Castara Road.

Scaled Pigeon
Uncommon. Mount St. Benedict, Arima Valley; often seen flying across valleys in Northern Range.

Band-tailed Pigeon
Rare. Seen in heights: El Tucuche, Las Lapas Trace.

Eared Dove
Difficult to observe in Trinidad but common in Tobago; Crown Point Airport region, Pigeon Point, Kilgwyn Swamp.

Common Ground-Dove
Uncommon and local. Caroni rice fields, Aripo Livestock Station (near cattle barns), Waller Field, along Churchill-Roosevelt Highway on wires between Arima and Cumuto Road.

Plain-breasted Ground-Dove
Upland areas in western Nariva Swamp and in the same areas as the previous species. Difficult to separate from it without good views (see ffrench for details).

White-tipped Dove
Uncommon and rarely seen in Trinidad, although call is heard throughout the day in Northern Range. Somewhat more common on Tobago; decidedly common at Grafton Estate at the 4 p.m. feeding. Prefers drier regions than Gray-fronted Dove.

Blue-and-yellow Macaw
Extirpated; last stronghold was Nariva Swamp. Feral flock in northern Tobago near Charlotteville reduced to a lone bird by late 1992.

Red-bellied Macaw
Locally abundant. Nariva Swamp/Waller Field, at times in flocks of more than 100. Frequents moriche and royal palms. Illegally hunted as cage bird (nest trees felled, young captured).

Green-rumped Parrotlet
Uncommon at higher elevations, more common in swamps and lowlands, singly or in small groups.

Lilac-tailed Parrotlet
Prefers higher elevations than previous species, especially rainforests; usually in flocks, sometimes large numbers. Inspect termitaria (termite structures) for silently feeding or nesting birds.

Scarlet-shouldered Parrotlet
Rare. Large flocks (40 birds) seen twice in recent years, once over Aripo Savannah (10 August) and once over Carapo, just west of Waller Field (18 March).

Blue-headed Parrot
Prefers higher elevations than Orange-winged. Common in Northern Range. Wing beats deeper and voice higher pitched than Orange-winged Parrot.

Yellow-crowned Parrot
Nariva Swamp, eastern side of Port-of-Spain, and vicinity of University of the West Indies campus in Curepe. Introduced. Mentioned here because of the confused nature of its taxonomy. Both the AOU and J.M. Forshaw (*Parrots of the World*, 3rd rev. ed., 1989) use the name 'Yellow-headed Parrot' to refer to *Amazona oratrix*, but the form (or species) thought to breed in Trinidad is the Yellow-crowned Parrot, *A. ochrocephala*, according to the AOU. Forshaw used the name 'Yellow-crowned Parrot' for *ochrocephala* and other races, including *oratrix*. It appears that the name 'Yellow-headed Parrot' pertains to the nominate form, *ochrocephala ochrocephala*.

Mangrove Cuckoo
In mangroves. Trinidad: Caroni Swamp. Tobago: Buccoo Marsh, Pigeon Point.

Dark-billed Cuckoo
Very rare, Caroni and Nariva Swamps.

Little Cuckoo
Uncommon and local. Waller Field, Caroni Swamp. Skulks in lower branches. Responds to pishing.

Greater Ani
Uncommon. Caroni, Nariva, and South Oropuche Swamps, Cacandee. Caroni boatmen should be able to find it. Usually found in family groups of 4-7.

Barn-Owl
Common, especially Caroni rice fields, Diego Martin and Chaguaramas areas.

Spectacled Owl
Rare. Most often found at the Asa Wright Nature Centre, along Las Lapas Trace, and in the Chaguaramas area of northwestern Trinidad.

Short-tailed Nighthawk (formerly Semicollared Nighthawk)
Apparently increasing. Identify by low, erratic, bat-like flight (associates with bats). Hunts over marshes and forests; often seen at twilight along Blanchisseuse Road near Arima and sometimes from veranda of Asa Wright Nature Centre.

Nacunda Nighthawk
Rare. Occasionally seen at dawn and dusk hawking flying insects over Waller Field.

Common Pauraque
Uncommon. Dawn and dusk at Waller Field and near fire tower on Mount St. Benedict.

Rufous Nightjar
Uncommon and local. Mount St. Benedict, Maracas Valley.

White-tailed Nightjar
Uncommon and usually silent. Dawn and dusk over Blanchisseuse Road near Arima, on the abandoned Waller Field runways at night. In Tobago: at Pigeon Point and Buccoo Marsh; roosts by day on mangrove roots.

Common Potoo
Not uncommon on either island but camouflage renders it hard to spot at its daytime roosts. Almost certain to be seen on boat trips in Caroni Swamp or night trips to Waller Field.

Oilbird
Rare and local, caves only. Best site is Dunstan Cave on the grounds of the Asa Wright Nature Centre or Oropuche and Cumaca caves with a local guide.

Chestnut-collared Swift
Uncommon. Flies with other species of swifts over foothills and in heights. Nests at the Asa Wright Nature Centre at the entrance to Dunstan Cave.

White-collared Swift
Rare but large and conspicuous, especially at low elevations just before rain showers; at other times in heights. Large flocks of migrants pass over at limits of visibility.

Chapman's Swift
Least common *Chaetura* swift. Difficult to identify even when seen from above. Best viewing spots are overlooks on Blanchisseuse Road between Textel entrance and Las Lapas Trace, Waller Field at dawn or dusk.

Short-tailed Swift
Chaetura species seen most often, especially at lower elevations and over urban areas.

Lesser Swallow-tailed Swift
Rare but easily identified. Most often reported over Mount St. Benedict and over Textel station on Blanchisseuse Road.

Fork-tailed Palm-Swift
Uncommon; frequents savannahs with moriche palms. Waller Field, Aripo Savannah, Nariva Swamp, Port-of-Spain near the Botanical Gardens.

White-tailed Sabrewing
Increasing since their nadir in early 1970s. Still extremely rare. Tobago only; trails in Main Ridge Forest Reserve.

Brown Violet-ear
Rare. A few individuals frequent estate house on L'Orange Road north of Ruiz Trace (Heights of Aripo); infrequent at Asa Wright Nature Centre. Often catches insects on the wing, a behavioural characteristic of possible identification value.

Green-throated Mango
Caroni Swamp, South Oropuche Swamp.

Blue-tailed Emerald
Drier areas from Mount St. Benedict west to the tip of the Chaguaramas Peninsula.

White-tailed Goldenthroat
Nariva Swamp, Caroni rice fields, Waller Field, Aripo Savanna, and other wet, low-lying savannahs.

Trogons
Can be distinguished by calls, all of which are ventriloquial, given with closed beak. Collared is unique – short, 3-5 notes, starts loud then diminishes in volume. White-tailed gives 10-20 notes; compare with Violaceous, with about same number of notes given twice as fast.

Collared Trogon
Common but inconspicuous and highly ventriloquial. Morne Bleu and other heights. The only trogon in Tobago, population increasing.

Blue-crowned Motmot
Uncommon in Trinidad, common and tame in Tobago. Look and listen (*'hoooooot'*) in Arena Forest. Frequents feeders at Asa Wright Nature Centre.

Ringed Kingfisher
Caroni rice fields, Waterloo, Nariva Swamp, Trincity Ponds.

American Pygmy Kingfisher
Cacandee, South Oropuche Swamp, Bush Bush Creek, Damier River. Uncommon and inconspicuous but widespread. Prefers small streams and tidal creeks, where it plunge-dives for minnows; also catches insects along streams.

Rufous-tailed Jacamar
Uncommon and elusive in Trinidad (Arena Forest, Mount St. Benedict), more often heard than seen. Common and tame in Tobago.

Channel-billed Toucan
Common but often feeds in treetops where difficult to spot; loud, raucous calls announce its presence. Northern Range, Arena Forest.

Crimson-crested Woodpecker
Rare. Prefers lowland forests with large, standing dead trees; sometimes seen in lower Arima Valley, Maridale Road.

Black-crested Antshrike
Uncommon. Found in most swamps and Waller Field seeps. Most active 1:30-3:30 p.m. when most other species are quiet.

Plain Antvireo
Uncommon. Forest heights along Blanchisseuse and Lalaja Roads.

Silvered Antbird
Most likely at Bush Bush Creek bridge on Mayaro Road and at northern end of bridge over Marianne River east of Blanchisseuse. Loud, long call rises up scale, then falls; responds quickly to taped playback.

White-bellied Antbird
Common but easily overlooked. Forages on ground in forests, Arena Forest. Woodpecker-like calls confirm its presence.

Black-faced Antthrush
Widespread but elusive at middle to high elevations in rainforests. Melancholy 4-6 note whistle can be heard throughout day. Easily attracted to whistled imitation of call.

Scaled Antpitta
Formerly considered to be extirpated, but call notes were heard in 1976 in the Matura Forest in southeastern Trinidad and two or more individuals were seen at the Asa Wright Nature Centre and vicinity and near Las Lapas Trace in 2001.

Flycatchers
Flycatchers fall into two groups – those easy to identify and those almost impossible to identify. Excluding unique Great Kiskadee, Fork-tailed Flycatcher, and the like, identification of members of this group should be based on physical characteristics in combination with call notes, especially for *Myiarchus* flycatchers. Many rare species seem more common in Trinidad's northwestern coastal scrub forests than elsewhere, except on islands between Trinidad and Venezuela (excluded from this guide).

Olive-striped Flycatcher
Heights of Aripo.

Mouse-colored Tyrannulet
Chaguaramas Peninsula.

Crested Doradito
Freshwater swamps bordering the Caroni and South Oropuche Swamps, all records June–September.

Northern Scrub-Flycatcher
Chaguaramas Peninsula.

Short-tailed Pygmy-Tyrant
Throughout Trinidad; rare along northeastern coast and in Arena Forest. Has nested at Asa Wright Nature Centre low in tree.

Spotted Tody-Flycatcher
South Oropuche and other swamps in southwestern Trinidad.

Yellow-olive Flycatcher
Rare. Asa Wright Nature Centre.

White-throated Spadebill
Trinidad: along the northeastern coast, hills around Salybia, Arena Forest, and throughout the Central Range. Tobago: Gilpin Trace. Sits motionless for extended periods of time.

Bran-colored Flycatcher
Uncommon; low-lying areas such as Waller Field.

Venezuelan Flycatcher
Tobago; Gilpin Trace.

Fork-tailed Flycatcher
Abundant in 'summer' season, scarce at other times; noted here because it roosts in enormous numbers in Caroni Swamp mangroves, flying in at twilight and leaving shortly after sunrise.

Sulphury Flycatcher
Uncommon; prefers moriche palms at Waller Field and the Aripo Savanna east of Cumuto. Mewing call is similar to that of Smooth-billed Ani.

White-winged Becard
Chaguaramas Peninsula. More common on Tobago.

Bearded Bellbird
Uncommon but usually within earshot anywhere in Northern Range except late August through October. Ventriloquial and difficult to spot. Accessible calling grounds on Asa Wright Nature Centre property, along Blanchisseuse Road, Ruiz Trace (Heights of Aripo).

Golden-headed Manakin
Trinidad only, where numerically the most abundant rainforest species. Leks are 10 m (40 ft.) above ground level. Highly ventriloquial; when its distinctive, rifle-ricochet call *'pit-tu'* is heard, look twice the height of perceived perch to spot bird.

Red-eyed ('Chivi') Vireo
Common but status uncertain; regarded by some as a separate species. Closely resembles Red-eyed Vireo in appearance, song, and behaviour.

White-winged Swallow
Uncommon, small numbers at Arena Reservoir, Waller Field, Aripo Livestock Station, Waterloo.

Blue-and-white Swallow
Mainly southwestern Trinidad north to Cacandee. Vagrant at Trincity Ponds. Hawks for insects over sugar cane fields. Roosts in mangroves in southern Caroni Swamp.

Southern Rough-winged Swallow
Common. Differs from Northern Rough-winged Swallow primarily by conspicuous apricot-coloured rump. Waller Field, Blanchisseuse, Piarco Water Treatment Plant.

'Tropical' House Wren
Abundant, particularly around human habitation; mentioned here because differs significantly enough for possible future split from the North American House Wren. Song and colouration of 'Tropical' House Wren is distinctly different from that of its northern counterparts. It is cinnamon, not grey-brown.

Long-billed Gnatwren
Common and widespread but elusive in understory, where dry, monotone trill announces presence.

Greater Bird-of-Paradise
Formerly resident, introduced from Aru Islands off New Guinea. Hard hit by Hurricane Flora in 1963. No sightings in recent years. Undoubtedly extirpated.

Thrushes
As a group much more active and vocal in evening than in morning. Most prefer middle to high elevations.

Orange-billed Nightingale-Thrush
Very rare and regarded as extremely secretive. Found in the highest points only – Heights of Aripo, El Tucuche, and Morne Bleu above Asa Wright Nature Centre along Blanchisseuse Road.

'North American' warblers
Many species have been seen from once to many times, often in strikingly dissimilar habitats than those utilized for nesting. Usually silent on wintering grounds. Species most likely to be seen include Yellow (perhaps Caribbean race), Northern Waterthrush (abundant), American Redstart (common), Blackpoll, Black-and-white, and Prothonotary (mangroves).

Common Yellowthroat
Probably very rare but status unclear because of similarity with the following species, from which it can be told by brown (not gray) crown, black mask extending well (not just very narrowly) above bill, and its pale belly (Masked is solidly yellow below).

Masked Yellowthroat
Rare but probably more common than records indicate. Inhabits same grassy and marshy habitats as Common Yellowthroat. Caroni rice fields, high grasses at entrance to Trincity Ponds. Responds to 'pishing'. See Common Yellowthroat.

Golden-crowned Warbler
Common but often overlooked in low to middle range of rainforest. Most frequently seen along Blanchisseuse Road above Asa Wright Nature Centre. Song weak, non-descript, of little value in locating bird.

Bicolored Conebill
Uncommon and local; mangroves. Most easily seen in mangroves along banks of canals in Caroni Swamp; also possible along Nariva River (Mayaro Road) and Bush Bush Creek. Chatters.

Red-crowned Ant-Tanager
Not uncommon but difficult to view, darting away rather than investigating pishing or owl calls. Along trail to Dunstan (Oilbird)

Cave at Asa Wright Nature Centre, also near wooden bridge just before Hollis Dam (north of Valencia).

Hepatic Tanager
Morne Bleu.

Blue-capped Tanager
Rare. Prefers higher elevations; Morne Bleu, Heights of Aripo.

Golden-rumped Euphonia
In hills above Mount St. Benedict. Most records probably refer to escaped cage birds.

Trinidad Euphonia
Rare and local because of trapping for use as cage bird. Regularly seen in Arima Valley, particularly at Simla. Elsewhere most easily seen in Central Range.

Speckled Tanager
Uncommon in small flocks. Prefers heights; Heights of Aripo, Morne Bleu, Andrews Trace.

Swallow-Tanager
Rare and seasonal. Flycatches from treetops in forest heights; Morne Bleu, above Asa Wright Nature Centre along Blanchisseuse Road, Las Lapas Trace, Andrews Trace, Heights of Aripo.

Seedeaters
Most species rare and decreasing rapidly. Trapping has reduced numbers of finches except for mediocre singers such as Blue-black and Black-faced Grassquits. Plum Mitan Road can be good for seedeaters.

Ruddy-breasted Seedeater
Aripo Livestock Station, grassy margins of runways at Crown Point Airport.

Sooty Grassquit
Morne Bleu, Heights of Aripo.

Saffron Finch
Common but local, especially in suburban areas from Pointe-à-Pierre south; sometimes seen in Port-of-Spain Botanical Gardens.

Red-capped Cardinal
Rare and elusive. Pointe-à-Pierre Wildfowl Trust, Caroni Swamp, Cacandee.

Streaked Saltator
Chaguaramas Peninsula (Teteron and Scotland Bays).

Dickcissel
Local. Nariva Swamp.

Giant Cowbird
Common but local; Aripo Livestock Station, Aripo Savannah, Waller Field. Large communal roost near Parlatuvier, Tobago. Closely resembles Common Grackle in size and flight style.

Moriche Oriole
Rarely seen. Aripo Savannah east of Cumuto, Waller Field, Nariva Swamp. Prefers moriche palms with fruiting figs nearby.

Common Waxbill
Introduced African species. Caroni rice fields, Trincity Ponds, central Caroni plains (Orange Grove).

ACCIDENTAL OR EXCEPTIONALLY RARE BIRD SPECIES

Recorded for Trinidad or Tobago but so infrequently as not to be expected.

Bulwer's Petrel
Cory's Shearwater
Manx Shearwater
Sooty Shearwater
Wilson's Storm-Petrel
White-tailed Tropicbird
Masked Booby
Northern Gannet
Grey Heron
Reddish Egret
Agami ('Chestnut-bellied') Heron
White Ibis
Jabiru
Wood Stork
Lesser Yellow-headed Vulture
Greater Flamingo
Horned Screamer
Snow Goose
Comb Duck
Muscovy Duck
Mallard
Northern Shoveler
Green-winged Teal
Southern Pochard
Ring-necked Duck
Snail Kite
Black-collared Hawk
Rufous Crab-Hawk
Swainson's Hawk
Northern Crested-Caracara
Aplomado Falcon
Orange-breasted Falcon
Ash-throated Crake
Paint-billed Crake
American Coot
Caribbean Coot
Sungrebe
Double-striped Thick-knee
Snowy Plover
Common Ringed Plover
American Oystercatcher
American Avocet
Common Greenshank
Spotted Redshank
Wood Sandpiper
Terek Sandpiper
Eskimo Curlew
Long-billed Curlew
Black-tailed Godwit
Marbled Godwit
Baird's Sandpiper
Ruff
Wilson's Phalarope
South Polar Skua
Franklin's Gull
Herring Gull
Lesser Black-backed Gull
Kelp Gull
Sabine's Gull
Caspian Tern
Black Noddy
White (Fairy) Tern
Scaled Dove
Monk Parakeet
Scarlet Macaw
Red-shouldered Macaw
Scarlet-shouldered Parrotlet
Yellow-crowned Parrot
Black-billed Cuckoo
Burrowing Owl
Common Nighthawk
Black Swift
Rufous-shafted Woodstar
Amazon Kingfisher
Scaled Antpitta
Crested Doradito
Small-billed Elaenia
Variegated Flycatcher
White Bellbird
Yellow-throated Vireo
Cliff Swallow
Veery
Gray-cheeked Thrush
White Wagtail
Golden-winged Warbler
Northern Parula

Chestnut-sided Warbler
Magnolia Warbler
Cape May Warbler
Black-throated Blue Warbler
Yellow-rumped Warbler
Black-throated Green Warbler
Blackburnian Warbler
Prairie Warbler
Bay-breasted Warbler
Ovenbird
Common Yellowthroat
Hooded Warbler
Canada Warbler
Scarlet Tanager
Slate-colored Seedeater
Variable Seedeater
Yellow-bellied Seedeater
Gray Seedeater
Lesson's Seedeater
Large-billed Seed-Finch
Orange-fronted Yellow-Finch
Grassland Yellow-Finch
Rose-breasted Grosbeak
Indigo Bunting
Bobolink
Red-winged Blackbird
Troupial
Baltimore Oriole
Red Siskin
House Sparrow

LIST OF BIRD SPECIES OF TRINIDAD AND TOBAGO

The following list contains the names of all species of birds accepted by the Trinidad and Tobago Rare Birds Committee as occurring or having occurred in or near the islands of Trinidad and Tobago. On 31 July 2004 the official list of species of Trinidad and Tobago stood at 457. For the latest official list of bird species visit http://www.geocities.com/ttrbc/ttrbcLst.html

i – introduced
m – spring and autumn migrant
r – resident
s – summer visitor
T – recorded only on Trinidad
To – recorded only on Tobago
v – vagrant
w – winter visitor
x – extirpated

1 – common
2 – uncommon
3 – rare

Tinamous – Tinamidae

rT2 ☐ Little Tinamou, *Crypturellus soui*

Grebes – Podicipedidae

r3 ☐ Least Grebe, *Tachybaptus dominicus*
r3 ☐ Pied-billed Grebe, *Podilymbus podiceps*

Petrels and Shearwaters – Procellariidae

v ☐ Bulwer's Petrel, *Bulweria bulwerii*
v ☐ Cory's Shearwater, *Calonectris diomedea*
v ☐ Greater Shearwater, *Puffinus gravis*
v ☐ Sooty Shearwater, *Puffinus griseus*
v ☐ Manx Shearwater, *Puffinus puffinus*
rTo3 ☐ Audubon's Shearwater, *Puffinus lherminieri*

Storm-Petrels – Hydrobatidae

mw3 ☐ Leach's Storm-Petrel, *Oceanodroma leucorhoa*

Tropicbirds – Phaethontidae

rTo3 ☐ White-tailed Tropicbird, *Phaethon lepturus*
mwTo3 ☐ Red-billed Tropicbird, *Phaethon aethereus*

Boobies – Sulidae

vTo ☐ Masked Booby, *Sula dactylatra*
r2 ☐ Brown Booby, *Sula leucogaster*
mwTo3 ☐ Red-footed Booby, *Sula sula*
v ☐ Northern Gannet, *Morus bassanus*

Pelicans – Pelicanidae

r1 ☐ Brown Pelican, *Pelicanus occidentalis*

Cormorants – Phalacrocoracidae

rsT2 ☐ Neotropic Cormorant, *Phalacrocorax brasilianus*

Anhingas – Anhingidae

r1 ☐ Anhinga, *Anhinga anhinga*

Frigatebirds – Fregatidae

r1 ☐ Magnificent Frigatebird, *Fregata magnificens*

Herons – Ardeidae

rT3 ☐ Pinnated Bittern, *Botaurus pinnatus*
rT3 ☐ Least Bittern, *Ixobrychus exilis*
rT3 ☐ Stripe-backed Bittern, *Ixobrychus involucris*
vT ☐ Rufescent Tiger-Heron, *Tigrisoma lineatum*
mr3 ☐ Great Blue Heron, *Ardea herodias*
vT ☐ Grey Heron, *Ardea cinerea*
vT ☐ Purple Heron, *Ardea purpurea*
w3 ☐ Cocoi Heron, *Ardea cocoi*
mw2 ☐ Great Egret, *Ardea alba*
v ☐ Little Egret, *Egretta garzetta*
v ☐ Western Reef-Egret, *Egretta gularis*
r1 ☐ Snowy Egret, *Egretta thula*
r1 ☐ Little Blue Heron, *Egretta caerulea*
r1 ☐ Tricolored Heron, *Egretta tricolor*
v ☐ Reddish Egret, *Egretta rufescens*
r1 ☐ Cattle Egret, *Bubulcus ibis*
r(?)w1 ☐ Green Heron, *Butorides virescens*
rT1 ☐ Striated Heron, *Butorides striatus*
vT ☐ Agami Heron, *Agamia agami*
r2 ☐ Black-crowned Night-Heron, *Nycticorax nycticorax*
r1 ☐ Yellow-crowned Night-Heron, *Nyctanassa violacea*
rT3 ☐ Boat-billed Heron, *Cochlearius cochlearius*

Ibises and Spoonbills – Threskiornithidae

vT ☐ White Ibis, *Eudocimus albus*
mrwT2 ☐ Scarlet Ibis, *Eudocimus ruber*
rT3 ☐ Glossy Ibis, *Plegadis falcinellus*
vT ☐ Roseate Spoonbill, *Ajaia ajaja*
vTo ☐ Eurasian Spoonbill, *Platalia leucorodia*

Storks – Ciconiidae

vT ☐ Jabiru, *Jabiru mycteria*
vT ☐ Wood Stork, *Mycteria americana*
vT ☐ Maguari Stork, *Ciconia maguari*

New World Vultures – Cathartidae

rT1 ☐ Black Vulture, *Coragyps atratus*

rT1 ☐ Turkey Vulture, *Cathartes aura*
vT3 ☐ King Vulture, *Sarcoramphus papa*

Flamingoes – Phoenicopteridae
vT ☐ Greater Flamingo, *Phoenicopterus ruber*

Screamers – Anhimidae
xT ☐ Horned Screamer, *Anhima cornuta*

Ducks and Geese – Anatidae
vT ☐ White-faced Whistling-Duck, *Dendrocygna viduata*
r2 ☐ Black-bellied Whistling-Duck, *Dendrocygna autumnalis*
r2 ☐ Fulvous Whistling-Duck, *Dendrocygna bicolor*
vT ☐ Snow Goose, *Chen caerulescens*
vT ☐ Comb Duck, *Sarkidiornis melanotos*
vT ☐ Muscovy Duck, *Cairina moschata*
w2 ☐ American Wigeon, *Anas americana*
mw2 ☐ Blue-winged Teal, *Anas discors*
wT3 ☐ Northern Shoveler, *Anas clypeata*
r2 ☐ White-cheeked Pintail, *Anas bahamensis*
vT ☐ Northern Pintail, *Anas acuta*
vT ☐ Green-winged Teal, *Anas crecca*
vT ☐ Southern Pochard, *Netta erythrophthalma*
vT ☐ Ring-necked Duck, *Aythya collaris*
vT ☐ Lesser Scaup, *Aythya affinis*
mw3 ☐ Masked Duck, *Nomonyx dominicus*

Hawks, Kites and Eagles – Accipitridae
mrw2 ☐ Osprey, *Pandion haliaetus*
rwT2 ☐ Gray-headed Kite, *Leptodon cayanensis*
vT ☐ Hook-billed Kite, *Chondrohierax uncinatus*
rs2 ☐ Swallow-tailed Kite, *Elanoides forficatus*
rT2 ☐ Pearl Kite, *Gampsonyx swainsonii*
rT2 ☐ White-tailed Kite, *Elanus leucurus*
vT ☐ Snail Kite, *Rostrhamus sociabilis*
rT2 ☐ Double-toothed Kite, *Harpagus bidentatus*
sT2 ☐ Plumbeous Kite, *Ictinia plumbea*
vT ☐ Black-collared Hawk, *Busarellus nigricollis*
mswT3 ☐ Long-winged Harrier, *Circus biffoni*
v ☐ Crane Hawk, *Geranospiza caerulescens*
rT1 ☐ White Hawk, *Leucopternis albicollis*
r1 ☐ Gray Hawk, *Asturina nitida*
vT ☐ Rufous Crab-Hawk, *Buteogallus aequinoctialis*
rT1 ☐ Common Black-Hawk, *Buteogallus anthracinus*
r3 ☐ Great Black-Hawk, *Buteogallus urubitinga*
rT2 ☐ Savanna Hawk, *Buteogallus meridionalis*
mr2 ☐ Broad-winged Hawk, *Buteo platypterus*
r2 ☐ Short-tailed Hawk, *Buteo brachyurus*
s3 ☐ Swainson's Hawk, *Buteo swainsoni*
rT3 ☐ White-tailed Hawk, *Buteo albicaudatus*

rT2 ☐ Zone-tailed Hawk, *Buteo albonotatus*
vT ☐ Black Hawk-Eagle, *Spizaetus tyrannus*
rT3 ☐ Ornate Hawk-Eagle, *Spizaetus ornatus*

Falcons and Caracaras – Falconidae
vT ☐ Northern Crested-Caracara, *Caracara cheriway*
r2 ☐ Yellow-headed Caracara, *Milvago chimachima*
v ☐ American Kestrel, *Falco sparverius*
vT ☐ Eurasian Kestrel, *Falco tinnunculus*
w2 ☐ Merlin, *Falco columbarius*
v ☐ Aplomado Falcon, *Falco femoralis*
rT2 ☐ Bat Falcon, *Falco rufigularis*
vT ☐ Orange-breasted Falcon, *Falco deiroleucus*
w2 ☐ Peregrine Falcon, *Falco peregrinus*

Guans and Chachalacas – Cracidae
rTo1 ☐ Rufous-vented Chachalaca, *Ortalis ruficauda*
rT3 ☐ Trinidad Piping-Guan, *Pipile pipile*

Gallinules, Rails and Crakes – Rallidae
rT3 ☐ Gray-breasted Crake, *Laterallus exilis*
rT1 ☐ Clapper Rail, *Rallus longirostris*
vT ☐ Rufous-necked Wood-Rail, *Aramides axillaris*
rT3 ☐ Gray-necked Wood-Rail, *Aramides cajanea*
mw2 ☐ Sora, *Porzana carolina*
rT2 ☐ Yellow-breasted Crake, *Porzana flaviventer*
v ☐ Ash-throated Crake, *Porzana albicollis*
v ☐ Paint-billed Crake, *Neocrex erythrops*
rT3 ☐ Spotted Rail, *Rallus maculatus*
r2 ☐ Purple Gallinule, *Porphyrula martinica*
rT3 ☐ Azure Gallinule, *Porphyrula flavirostris*
r1 ☐ Common Moorhen, *Gallinula chloropus*
v ☐ American Coot, *Fulica americana*
v ☐ Caribbean Coot, *Fulica caribaea*

Sungrebes – Heliornithidae
vT ☐ Sungrebe, *Heliornis fulica*

Limpkins – Aramidae
rT3 ☐ Limpkin, *Aramus guarauna*

Thick-knees – Burhinidae
vT ☐ Double-striped Thick-knee, *Burhinus bistriatus*

Plovers and Lapwings – Charadriidae
vT ☐ Pied Plover, *Hoploxypterus cayanus*
r1 ☐ Southern Lapwing, *Vanellus chilensis*
mr1 ☐ Black-bellied (Grey) Plover, *Pluvialis squatarola*
m3 ☐ American Golden-Plover, *Pluvialis dominica*
ms1 ☐ Collared Plover, *Charadrius collaris*

r3T ☐ Wilson's Plover, *Charadrius wilsonia*
vT ☐ Common Ringed Plover, *Charadrius hiaticula*
mrw2 ☐ Semipalmated Plover, *Charadrius semipalmatus*
wT3 ☐ Killdeer, *Charadrius vociferus*

Oystercatchers – Haematopodidae
v ☐ American Oystercatcher, *Haematopus palliatus*

Stilts and Avocets – Recurvirostridae
mwT3 ☐ Black-necked Stilt, *Himantopus mexicanus*
v ☐ American Avocet, *Recurvirostra americana*

Jacanas – Jacanidae
r1 ☐ Wattled Jacana, *Jacana jacana*

Sandpipers – Scolopacidae
mr2 ☐ Greater Yellowlegs, *Tringa melanoleuca*
mr2 ☐ Lesser Yellowlegs, *Tringa flavipes*
v ☐ Spotted Redshank, *Tringa erythropus*
v ☐ Wood Sandpiper, *Tringa glareola*
mw2 ☐ Solitary Sandpiper, *Tringa solitaria*
mr1 ☐ Willet, *Catoptrophorus semipalmatus*
mrw1 ☐ Spotted Sandpiper, *Actitis macularia*
vT ☐ Terek Sandpiper, *Xenus cinereus*
mT3 ☐ Upland Sandpiper, *Bartramia longicauda*
x ☐ Eskimo Curlew, *Numenius borealis*
mr3 ☐ Whimbrel, *Numenius phaeopus*
v ☐ Long-billed Curlew, *Numenius americanus*
vT ☐ Black-tailed Godwit, *Limosa limosa*
mT3 ☐ Hudsonian Godwit, *Limosa haemastica*
vT ☐ Marbled Godwit, *Limosa fedoa*
r3 ☐ Ruddy Turnstone, *Arenaria interpres*
mT3 ☐ Red Knot, *Calidris canutus*
mrw2 ☐ Sanderling, *Calidris alba*
v ☐ Curlew Sandpiper, *Calidris ferruginea*
mrw2 ☐ Semipalmated Sandpiper, *Calidris pusilla*
mrw2 ☐ Western Sandpiper, *Calidris mauri*
m3 ☐ Least Sandpiper, *Calidris minutilla*
m3 ☐ White-rumped Sandpiper, *Calidris fuscicollis*
vT ☐ Baird's Sandpiper, *Calidris bairdii*
m2 ☐ Pectoral Sandpiper, *Calidris melanotos*
m2 ☐ Stilt Sandpiper, *Calidris himantopus*
v ☐ Buff-breasted Sandpiper, *Tryngites subruficollis*
v ☐ Ruff, *Philomachus pugnax*
mrw2 ☐ Short-billed Dowitcher, *Limnodromus griseus*
mrw2 ☐ Wilson's Snipe, *Gallinago delicata*
v ☐ Wilson's Phalarope, *Phalaropus tricolor*

Skuas, Gulls and Terns – Laridae
v ☐ South Polar Skua, *Catharacta maccormicki*

v ☐ Pomarine Jaegar (Pomarine Skua), *Stercorcarius pomarinus*
sT3 ☐ Parasitic Jaegar (Arctic Skua), *Stercorcarius parasiticus*
rms2 ☐ Laughing Gull, *Larus atricilla*
vT ☐ Franklin's Gull, *Larus pipixcan*
v ☐ Black-headed Gull, *Larus ridibundus*
v ☐ Ring-billed Gull, *Larus delawarensis*
v ☐ Herring Gull, *Larus argentatus*
v ☐ Lesser Black-backed Gull, *Larus fuscus*
vT ☐ Kelp Gull, *Larus dominicanus*
v ☐ Black-legged Kittiwake, *Rissa tridactyla*
ms2 ☐ Gull-billed Tern, *Sterna nilotica*
v ☐ Caspian Tern, *Sterna caspia*
r2 ☐ Royal Tern, *Sterna maxima*
sw3 ☐ Sandwich Tern, *Sterna sandvicensis*
sw2 ☐ Roseate Tern, *Sterna dougallii*
rs2 ☐ Common Tern, *Sterna hirundo*
r3 ☐ Least Tern, *Sterna albifrons*
msT2 ☐ Yellow-billed Tern, *Sterna superciliaris*
sTo2 ☐ Bridled Tern, *Sterna anaethetus*
sTo1 ☐ Sooty Tern, *Sterna fuscata*
msT2 ☐ Large-billed Tern, *Phaetusa simplex*
mwT2 ☐ Black Tern, *Chlidonias niger*
mrs1 ☐ Brown Noddy, *Anous stolidus*
vTo ☐ White (Fairy) Tern, *Gygis alba*
rmsT2 ☐ Black Skimmer, *Rynchops niger*

Pigeons and Doves – Columbidae

irT2 ☐ Feral (Rock) Pigeon, *Columba livia*
rT2 ☐ Pale-vented Pigeon, *Columba cayennensis*
rT2 ☐ Scaled Pigeon, *Columba speciosa*
rT3 ☐ Band-tailed Pigeon, *Columba fasciata*
r2 ☐ Eared Dove, *Zenaida auriculata*
vT ☐ Scaled Dove, *Scardafella squammata*
rT3 ☐ Common Ground-Dove, *Columbina passerina*
rT3 ☐ Plain-breasted Ground-Dove, *Columbina minuta*
r1 ☐ Ruddy Ground-Dove, *Columbina talpacoti*
rT3 ☐ Blue Ground-Dove, *Claravis pretiosa*
r2 ☐ White-tipped Dove, *Leptotila verreauxi*
rT2 ☐ Gray-fronted Dove, *Leptotila rufaxilla*
r3 ☐ Lined Quail-Dove, *Geotrygon linearis*
rT3 ☐ Ruddy Quail-Dove, *Geotrygon montana*

Parrots and Macaws – Psittacidae

xT ☐ Scarlet Macaw, *Ara macao*
xT ☐ Blue-and-yellow Macaw, *Ara ararauna*
rT3 ☐ Red-bellied Macaw, *Ara manilata*
r1 ☐ Green-rumped Parrotlet, *Forpus passerinus*
rT2 ☐ Lilac-tailed Parrotlet, *Touit batavica*
vT ☐ Scarlet-shouldered Parrotlet, *Touit huetii*
rT1 ☐ Blue-headed Parrot, *Pionus menstruus*

r1 ☐ Orange-winged Parrot, *Amazona amazonica*
vT ☐ Yellow-crowned Parrot, *Amazona ochrocephala*

New World Cuckoos – Cuculidae

vT ☐ Black-billed Cuckoo, *Coccyzus erythropthalmus*
mT3 ☐ Yellow-billed Cuckoo, *Coccyzus americanus*
r3 ☐ Mangrove Cuckoo, *Coccyzus minor*
sT3 ☐ Dark-billed Cuckoo, *Coccyzus melacoryphus*
rT2 ☐ Squirrel Cuckoo, *Piaya cayana*
rT2 ☐ Little Cuckoo, *Piaya minuta*
rT2 ☐ Striped Cuckoo, *Tapera naevia*
rT2 ☐ Greater Ani, *Crotophaga major*
r1 ☐ Smooth billed Ani, *Crotophaga ani*

Barn Owls – Tytonidae

r3 ☐ Barn Owl, *Tyto alba*

Owls – Strigidae

rT1 ☐ Tropical Screech-Owl, *Otus choliba*
rT3 ☐ Spectacled Owl, *Pulsatrix perspicillata*
rT1 ☐ Ferruginous Pygmy-Owl, *Glaucidium brasilianum*
vT ☐ Burrowing Owl, *Athene cunicularia*
rT3 ☐ Mottled Owl, *Ciccaba virgata*
vT ☐ Short-eared Owl, *Asio flammeus*
rTo3 ☐ Striped Owl, *Pseudoscops clamator*

Nightjars and Nighthawks – Caprimulgidae

rT2 ☐ Short-tailed Nighthawk, *Lurocalis semitorquatus*
mr2 ☐ Lesser Nighthawk, *Chordeiles acutipennis*
rT2 ☐ Common Pauraque, *Nyctidromus albicollis*
rT3 ☐ Rufous Nightjar, *Caprilmulgus rufus*
r2 ☐ White-tailed Nightjar, *Caprilmulgus cayennensis*
m3 ☐ Nacunda Nighthawk, *Podager nacunda*

Potoos – Nyctibiidae

r2 ☐ Common Potoo, *Nyctibius griseus*

Oilbirds – Steatornithidae

rT3 ☐ Oilbird, *Steatornis caripensis*

Swifts – Apodidae

rT3 ☐ Chestnut-collared Swift, *Streptoprocne rutila*
mw2 ☐ White-collared Swift, *Streptococne zonaris*
rT3 ☐ Chapman's Swift, *Chaetura chapmani*
r1 ☐ Short-tailed Swift, *Chaetura brachyura*
rT2 ☐ Band-rumped Swift, *Chaetura spinicauda*
r2 ☐ Gray-rumped Swift, *Chaetura cinereiventris*
rT3 ☐ Fork-tailed Palm-Swift, *Reinarda squamata*
rT3 ☐ Lesser Swallow-tailed Swift, *Panyptila cayennensis*

Hummingbirds – Trochilidae

r1 ☐ Rufous-breasted Hermit, *Glaucis hirsuta*

rT1 ☐ Green Hermit, *Phaethornis guy*
rT3 ☐ Little Hermit, *Phaethornis longuemareus*
rTo3 ☐ White-tailed Sabrewing, *Campylopterus ensipennis*
r2 ☐ White-necked Jacobin, *Florisuga mellivora*
v ☐ Brown Violet-ear, *Colibri delphinae*
rT3 ☐ Green-throated Mango, *Anthracothorax viridigula*
r1 ☐ Black-throated Mango, *Anthracothorax nigricollis*
rT3 ☐ White-tailed Goldenthroat, *Polytmus guainumbi*
msw2 ☐ Ruby-topaz Hummingbird, *Chrysolampis mosquitus*
rT2 ☐ Tufted Coquette, *Lophornis ornata*
r1 ☐ Blue-chinned Sapphire, *Chlorestes notatus*
rT3 ☐ Blue-tailed Emerald, *Chlorostilbon mellisugus*
rT1 ☐ White-chested Emerald, *Amazilia chionopectus*
r1 ☐ Copper-rumped Hummingbird, *Amazilia tobaci*
mrw2 ☐ Long-billed Starthroat, *Heliomaster longirostris*
vT ☐ Rufous-shafted Woodstar, *Chaetocercus jourdanii*

Trogons – Trogonidae

rT1 ☐ White-tailed Trogon, *Trogon viridis*
rT1 ☐ Violaceous Trogon, *Trogon violaceus*
r2 ☐ Collared Trogon, *Trogon collaris*

Motmots – Momotidae

r1 ☐ Blue-crowned Motmot, *Momotus momota*

Kingfishers – Alcedinidae

msT3 ☐ Ringed Kingfisher, *Ceryle torquata*
w3 ☐ Belted Kingfisher, *Ceryle alcyon*
vT ☐ Amazon Kingfisher, *Chloroceryle amazona*
r2 ☐ Green Kingfisher, *Chloroceryle americana*
rT2 ☐ American Pygmy Kingfisher, *Chloroceryle aenea*

Jacamars – Galbulidae

r1 ☐ Rufous-tailed Jacamar, *Galbula ruficauda*

Toucans – Ramphastidae

rT1 ☐ Channel-billed Toucan, *Ramphastos vitellinus*

Woodpeckers – Picidae

rTo1 ☐ Red-crowned Woodpecker, *Melanerpes rubricapillus*
r2 ☐ Red-rumped Woodpecker, *Veniliornis kirkii*
r1 ☐ Golden-olive Woodpecker,*Piculus rubiginosus*
rT1 ☐ Chestnut Woodpecker, *Celeus elegans*
rT1 ☐ Lineated Woodpecker, *Dryocopus lineatus*
rT3 ☐ Crimson-crested Woodpecker, *Campephilus melanoleucos*

Ovenbirds – Furnariidae

rT2 ☐ Pale-breasted Spinetail, *Synallaxis albescens*
r2 ☐ Stripe-breasted Spinetail, *Synallaxis cinnamomea*
rT2 ☐ Yellow-chinned Spinetail, *Certhiaxis cinnamomea*

r2 ☐ Streaked Xenops, *Xenops rutilans*
r2 ☐ Gray-throated Leaftosser, *Sclerurus albigularis*

Woodcreepers – Dendrocolaptidae
r2 ☐ Plain-brown Woodcreeper, *Dendrocincla fuliginosa*
rTo2 ☐ Olivaceous Woodcreeper, *Sittasomus griseicapillus*
rT3 ☐ Straight-billed Woodcreeper, *Xiphorhynchus picus*
r2 ☐ Cocoa Woodcreeper, *Xiphorhynchus susurrans*
rT3 ☐ Streak-headed Woodcreeper, *Lepidocolaptes souleyetii*

Antbirds – Thamnophilidae
rT2 ☐ Great Antshrike, *Taraba major*
rT3 ☐ Black-crested Antshrike, *Sakesphorus canadensis*
r1 ☐ Barred Antshrike, *Thamnophilus doliatus*
r2 ☐ Plain Antvireo, *Dysithamnus mentalis*
rT2 ☐ White-flanked Antwren, *Myrmotherula axillaris*
rTo2 ☐ Northern White-fringed Antwren, *Formicivora grisea*
rT3 ☐ Silvered Antbird, *Sclateria naevia*
rT2 ☐ White-bellied Antbird, *Myrmeciza longipes*

Antthrushes and Antpittas – Formicariidae
rT2 ☐ Black-faced Antthrush, *Formicarius analis*
vT ☐ Scaled Antpitta, *Grallaria guatimalensis*

Flycatchers – Tyrannidae
rT3 ☐ Southern Beardless-Tyrannulet, *Camptostoma obsoletum*
msT3 ☐ Mouse-colored Tyrannulet, *Phaeomyias murina*
rT2 ☐ Forest Elaenia, *Myiopagis gaimardii*
r1 ☐ Yellow-bellied Elaenia, *Elaenia flavogaster*
msT3 ☐ Lesser Elaenia, *Elaenia chiriquensis*
vT ☐ Small-billed Elaenia, *Elaenia parvirostris*
vT ☐ Slaty Elaenia, *Elaenia strepera*
rT3 ☐ Olive-striped Flycatcher, *Mionectes olivaceus*
r2 ☐ Ochre-bellied Flycatcher, *Mionectes oleagineus*
rT2 ☐ Slaty-capped Flycatcher, *Leptopogon superciliaris*
rT3 ☐ Northern Scrub-Flycatcher, *Sublegatus arenarum*
rT3 ☐ Short-tailed Pygmy-Tyrant, *Myiornis ecaudatus*
vT ☐ Crested Doradito, *Pseudocolopteryx sclateri*
rT3 ☐ Spotted Tody-Flycatcher, *Todirostrum maculatum*
rT2 ☐ Yellow-olive Flycatcher, *Tolmomyias sulphurescens*
r1 ☐ Ochre-lored Flatbill, *Tolmonyias flavivrentris*
r3 ☐ White-throated Spadebill, *Platyrinchus mystaceus*
rT3 ☐ Bran-colored Flycatcher, *Myiophobus fasciatus*
rT2 ☐ Euler's Flycatcher, *Lathrotriccus euleri*
rT3 ☐ Fuscous Flycatcher, *Cnemotriccus fuscatus*
wT2 ☐ Olive-sided Flycatcher, *Contopus cooperi*
rT2 ☐ Tropical Pewee, *Contopus cinereus*
rT1 ☐ Pied Water-Tyrant, *Fluvicola pica*
rT1 ☐ White-headed Marsh-Tyrant, *Arundinicola leucocephala*
rT3 ☐ Bright-rumped Attila, *Attila spadiceus*

rT2 ☐ Dusky-capped Flycatcher, *Myiarchus tuberculifer*
vT ☐ Swainson's Flycatcher, *Myiarchus swainsoni*
r2 ☐ Brown-crested Flycatcher, *Myiarchus tyrannulus*
rTo3 ☐ Venezuelan Flycatcher, *Myiarchus venezuelensis*
rT1 ☐ Great Kiskadee, *Pitangus sulphuratus*
rT1 ☐ Boat-billed Flycatcher, *Megarynchus pitangua*
r2 ☐ Streaked Flycatcher, *Myiodynastes maculatus*
sw2 ☐ Piratic Flycatcher, *Legatus leucophaius*
vT ☐ Variegated Flycatcher, *Empidonomus varius*
rT3 ☐ Sulphury Flycatcher, *Tyrannopsis sulphurea*
r1 ☐ Tropical Kingbird, *Tyrannus melancholicus*
mrw2 ☐ Gray Kingbird, *Tyrannus dominicensis*
sw1 ☐ Fork-tailed Flycatcher, *Tyrannus savana*
rT3 ☐ White-winged Becard, *Pachyramphus polychopterus*
rT2 ☐ Black-tailed Tityra, *Tityra cayana*

Bellbirds – Cotingidae

vT ☐ White Bellbird, *Procnias alba*
rT2 ☐ Bearded Bellbird, *Procnias averano*

Manakins – Pipridae

rT1 ☐ White-bearded Manakin, *Manacus manacus*
rTo2 ☐ Blue-backed Manakin, *Chiroxiphia pareola*
rT1 ☐ Golden-headed Manakin, *Pipra erythrocephala*

Vireos – Vireonidae

v ☐ White-eyed Vireo, *Vireo griseus*
v ☐ Yellow-throated Vireo, *Vireo flavifrons*
mrs2 ☐ Red-eyed Vireo, *Vireo olivaceus*
mT3 ☐ Black-whiskered Vireo, *Vireo altiloquus*
rTo2 ☐ Scrub Greenlet, *Hylophilus flavipes*
rT1 ☐ Golden-fronted Greenlet, *Hylophilus aurantiifrons*
rT1 ☐ Rufous-browed Peppershrike, *Cyclarhis gujanensis*

Swallows and Martins – Hirundinidae

swTo2 ☐ Caribbean Martin, *Progne dominicensis*
rT2 ☐ Gray-breasted Martin, *Progne chalybea*
rT3 ☐ White-winged Swallow, *Tachycineta albiventer*
sT2v ☐ Blue-and-white Swallow, *Pygochelidon cyanoleuca*
r2v ☐ Southern Rough-winged Swallow, *Stelgidopteryx ruficollis*
v ☐ Bank Swallow (Sand Martin), *Riparia riparia*
v ☐ Cliff Swallow, *Petrochelidon pyrrhonota*
mw2 ☐ Barn Swallow, *Hirundo rustica*

Wrens – Troglodytidae

r2 ☐ Rufous-breasted Wren, *Thryothorus rutilus*
r1 ☐ House Wren, *Troglodytes aedon*

Gnatwrens – Sylviidae

rT2 ☐ Long-billed Gnatwren, *Ramphocaenus melanurus*

Thrushes – Turdidae

rT3 ☐ Orange-billed Nightingale-Thrush, *Catharus aurantiirostris*
vT ☐ Veery, *Catharus fuscescens*
vT ☐ Gray-cheeked Thrush, *Catharus minimus*
r2 ☐ Yellow-legged Thrush, *Platycichla flavipes*
rT1 ☐ Cocoa Thrush, *Turdus fumigatus*
r1 ☐ Bare-eyed Thrush, *Turdus nudigenis*
r2 ☐ White-necked Thrush, *Turdus albicollis*

Mockingbirds – Mimidae

r2 ☐ Tropical Mockingbird, *Mimus gilvus*

Wagtails – Motacillidae

vT ☐ White Wagtail, *Motacilla alba*

Wood Warblers – Parulidae

v ☐ Golden-winged Warbler, *Vermivora chrysoptera*
v ☐ Northern Parula, *Parula americana*
rT2 ☐ Tropical Parula, *Parula pitiayumi*
w2 ☐ Yellow Warbler, *Dendroica petechia*
v ☐ Chestnut-sided Warbler, *Dendroica pensylvanica*
v ☐ Magnolia Warbler, *Dendroica magnolia*
v ☐ Cape May Warbler, *Dendroica tigrina*
v ☐ Black-throated Blue Warbler *Dendroica caerulescens*
v ☐ Yellow-rumped Warbler, *Dendroica coronata*
v ☐ Black-throated Green Warbler, *Dendroica virens*
v ☐ Blackburnian Warbler, *Dendroica fusca*
v ☐ Prairie Warbler, *Dendroica discolor*
v ☐ Bay-breasted Warbler, *Dendroica castanea*
v ☐ Blackpoll Warbler, *Dendroica striata*
v ☐ Cerulean Warbler, *Dendroica cerulea*
vT ☐ Black-and-white Warbler, *Mniotilta varia*
w2 ☐ American Redstart, *Setophaga ruticilla*
w3 ☐ Prothonotary Warbler, *Protonotaria citrea*
v ☐ Ovenbird, *Seiurus aurocapillus*
mw2 ☐ Northern Waterthrush, *Seiurus noveboracensis*
v ☐ Common Yellowthroat, *Geothlypis trichas*
rT2 ☐ Masked Yellowthroat, *Geothlypis aequinoctialis*
v ☐ Hooded Warbler, *Wilsonia citrina*
v ☐ Canada Warbler, *Wilsonia canadensis*
r2 ☐ Golden-crowned Warbler, *Basileuterus culicivorus*

Bananaquits – Coerebidae

r1 ☐ Bananaquit, *Coereba flaveola*

Tanagers – Thraupidae

rT3 ☐ Bicolored Conebill, *Conirostrum bicolor*
rT2 ☐ White-shouldered Tanager, *Tachyphonus luctuosus*
r1 ☐ White-lined Tanager, *Tachyphonus rufus*
rT3 ☐ Red-crowned Ant-Tanager, *Habia rubica*

rT3 ☐ Hepatic Tanager, *Piranga flava*
wT3 ☐ Summer Tanager, *Piranga rubra*
vT ☐ Scarlet Tanager, *Piranga olivacea*
rT1 ☐ Silver-beaked Tanager, *Ramphocelus carbo*
r1 ☐ Blue-gray Tanager, *Thraupis episcopus*
r1 ☐ Palm Tanager, *Thraupis palmarum*
rT3 ☐ Blue-capped Tanager, *Thraupis cyanocephala*
r1 ☐ Violaceous Euphonia, *Euphonia violacea*
rT3 ☐ Trinidad Euphonia, *Euphonia trinitatis*
v ☐ Golden-rumped Euphonia, *Euphonia cyanocephala*
rT3 ☐ Speckled Tanager, *Tangara guttata*
rT2 ☐ Turquoise Tanager, *Tangara mexicana*
rT1 ☐ Bay-headed Tanager, *Tangara gyrola*
rT2 ☐ Blue Dacnis, *Dacnis cayana*
rT2 ☐ Green Honeycreeper, *Chlorophanes spiza*
r1 ☐ Purple Honeycreeper, *Cyanerpes caeruleus*
r2 ☐ Red-legged Honeycreeper, *Cyanerpes cyaneus*
swT3 ☐ Swallow-Tanager, *Tersina viridis*

Grassquits, Seedeaters and Finches – Emberizidae

r1 ☐ Blue-black Grassquit, *Volatinia jacarina*
rT3 ☐ Sooty Grassquit, *Tiaris fuliginosa*
v ☐ Slate-colored Seedeater, *Sporophila schistacea*
v ☐ Gray Seedeater, *Sporophila intermedia*
vTo ☐ Variable Seedeater, *Sporophila americana*
v ☐ Yellow-bellied Seedeater, *Sporophila nigricollis*
v ☐ Lesson's Seedeater, *Sporophila bouvronides*
r3 ☐ Ruddy-breasted Seedeater, *Sporophila minuta*
v ☐ Large-billed Seed-Finch, *Oryzoborus crassirostris*
v ☐ Lesser Seed-Finch, *Oryzborus angolensis*
rTo1 ☐ Black-faced Grassquit, *Tiaris bicolor*
rT2 ☐ Saffron Finch, *Sicalis flaveola*
vT ☐ Grassland Yellow-Finch, *Sicalis luteola*
v ☐ Orange-fronted Yellow-Finch, *Sicalis columbiana*
rT3 ☐ Red-capped Cardinal, *Paroaria gularis*

Saltators and Buntings – Cardinalidae

vT ☐ Streaked Saltator, *Saltator striatipectus*
rT2 ☐ Grayish Saltator, *Saltator coerulescens*
vv ☐ Rose-breasted Grosbeak, *Pheucticius ludovicianus*
v ☐ Indigo Bunting, *Passerina cyanea*
mwT1 ☐ Dickcissel, *Spiza americana*

Cowbirds, New World Blackbirds and Orioles – Icteridae

vT ☐ Bobolink, *Dolichonyx oryzivorus*
v ☐ Red-winged Blackbird, *Agelaius phoeniceus*
r2 ☐ Red-breasted Blackbird, *Sturnella militaris*
rT2 ☐ Yellow-hooded Blackbird, *Agelaius icterocephalus*
r1 ☐ Carib Grackle, *Quiscalus lugubris*
r1 ☐ Shiny Cowbird, *Molothrus bonariensis*

r2 ☐ Giant Cowbird, *Molothrus oryzivora*
vT ☐ Moriche Oriole, *Icterus chrysocephalus*
rT2 ☐ Yellow Oriole, *Icterus nigrogularis*
v ☐ Baltimore Oriole, *Icterus galbula*
rT2 ☐ Yellow-rumped Cacique, *Cacicus cela*
r1 ☐ Crested Oropendola, *Psarocolius decumanus*

Siskins – Carduelinae

xT ☐ Red Siskin, *Carduelis cucullata*

Estrildidae

iT ☐ Common Waxbill, *Estrilda astrild*

Black-tailed Tityra

LAWS PERTAINING TO BIRDS IN TRINIDAD AND TOBAGO

The following birds are classified as game. Daily bag limits are 30 ducks and five night-herons per person.

Neotropic Cormorant
Double-crested Cormorant (designated although as yet never recorded from either island!)
Great Blue Heron
Cocoi Heron
Black-crowned Night-Heron
Yellow-crowned Night-Heron
Boat-billed Heron
Black-bellied Whistling-Duck
All ducks except White-cheeked Pintail and Muscovy
Black Vulture
All rallids (coots, rails, etc.)
All plovers
All scolopacids (snipes, curlews, sandpipers, yellowlegs, godwits, etc.)
All pigeons
Orange-winged Parrot

The following species may be captured or kept in captivity provided the cage area is not less than one cubic foot per captive bird:

Green-rumped Parrotlet
Trinidad Euphonia
Violaceous Euphonia
Gray Seedeater
Lesson's Seedeater
Yellow-bellied Seedeater
Lesser Seed-Finch

The following species are declared vermin:

Crested Oropendola
Orange-winged Parrot
Rufous-vented Chachalaca

BRIEF SUMMARY OF TRINIDAD AND TOBAGO'S BIOTA

Based on Convention on Biological Diversity, 'Interim First National Report (revised) to the Convention on Biological Diversity (2000)' http://www.biodiv.org/doc/world/tt/tt-nr-01-en.pdf

LARGE PLANT FAMILIES

Large families in terms of species numbers in Trinidad and Tobago (approximately 55% of the total number of vascular plants):

Family	Genera	Species
Ferns & allies	66	310
Grasses	74	214
Legumes	75	202
Orchids	68	190
Sedges	22	111
'Rubes' (Rubiaceae)	48	97
Melastomes	22	95
Composites	45	86
Euphorbs	27	82

FRESHWATER FISH (approximately 45 species and excluding sea run and peripheral species)

Twenty-one families of freshwater fish occur in Trinidad, in few orders, four of which are marine families with freshwater representatives. Tobago, in contrast, only has seven families, four of which are marine families with freshwater representatives

MARINE FISH

Although there have been no recent taxonomic reviews of the marine fish of Trinidad and Tobago, it is possible to project an ichthyofauna of 400-500 species in several dozens orders and families.

AMPHIBIA

Amphibians constitute the smallest group of vertebrates. In Trinidad there are about 30 species in nine families, all in a single order, while Tobago has only about 10 species in fewer families. All are of the anuran order, frogs or toads, there being no salamanders or caecilians (legless amphibians). One frog species might be endemic (*Phyllodytes auratus* – Golden Tree Frog).

REPTILES (approximately 85 species, including marine turtles)

About 40 species of snakes in six families are found in Trinidad. In Tobago there are fewer families and numbers of species. About 25 species of lizards in five families are found in Trinidad and Tobago. They vary considerably in size, habits, and distribution. Some are arboreal, a few terrestrial, and a few burrowing. The rest of the reptilian fauna includes turtles, terrapins and tortoises, and a single crocodilian, the caiman or alligator. Only the tortoise is terrestrial. Two species of tortoise occur; one native and one introduced, but both are extremely rare in the wild.

BIRDS (457 species, of which about 250 breed locally)

The avifauna of Trinidad and Tobago is extremely well documented in the technical and popular literature. Birds constitute the largest group of vertebrates. 66 families in 20 orders are represented in Trinidad and Tobago.

The dominant order, as in many other parts of the world, is the Passeriformes, accounting for almost a third of the families represented. About 400 species have been recorded in Trinidad. As with other vertebrate groups, there are substantially fewer species in Tobago than in Trinidad. About 170 species have been recorded for Tobago.

Certain general features of the avifauna may be noted. A substantial proportion are resident and breeding but there are also migrants from North America, a few from South America, seabirds typical of the Caribbean biogeographic province, as well as oceanic species, occasional visitors, and strays.

MAMMALS (about 95 species including marine mammals)

Mammals constitute the final vertebrate group of the terrestrial biota and are well documented. Nine orders and about 27 families are represented and all are typical of the adjacent mainland and the wider neotropics. There are approximately 100 indigenous species but bats and rodents predominate. Bats alone account for more than half of the mammalian fauna. The rest of the mammalian group includes marsupials, edentates, a single armadillo, several rodents, primates, a few carnivores, deer, and the manatee.

OTHER GROUPS

Insects and arachnids dominate the terrestrial fauna. Unfortunately, knowledge of the various groups is very uneven. Orders such as the Coleoptera (beetles), Lepidoptera (butterflies and moths), and Diptera (flies) are relatively large but there are immense gaps. The single volume on lepidopterans lists 617 species of butterflies. Most of the hesperiidae (doctor butterflies) or moths are excluded, except for hawk moths. The total number of lepidopteran species is therefore likely to be much higher.

Some of the flies, for example mosquitoes and sandflies, are particularly well known, owing to their importance as pests and vectors of disease. Pests are generally better known as an amorphous grouping rather than as members of one or more insect order.

In contrast to insects, arachnids are very poorly documented. This group is dominated by spiders, ticks, and mites but includes scorpions and a few relatives. The remaining terrestrial invertebrate fauna is poorly documented.

CHECKLIST OF MAMMALS OF TRINIDAD & TOBAGO

Eisenberg, J. F. 1989. *Mammals of the Neotropics. Vol. 1. The Northern Neotropics.* University of Chicago Press, Chicago. 449 pp.

B = resident
I = introduced
X = extinct

ORDER/Family/*Species*	**Trinidad**	**Tobago**
MARSUPIALA		
Didelphidae		
Wooly Opossum		
Caluromys philander	B	–
Water Opossum		
Chironectes minimus	B	–
Common Opossum		
Didelphis marsupialis	B	B
Dusky Mouse Opossum		
Marmosa fuscata	B	–
Rat Opossum		
Marmosa murina	B?	B
Robinson's Opossum		
Marmosa robinsoni	B	–
EDENTATA		
Myrmecophagidae		
Lesser Anteater		
Tamandua tetradactyla	B	–
Silky Anteater		
Cyclopes didactylus	B	–
Dasypodidae		
Nine-banded Armadillo		
Dasypus novemcinctus	B	B
CHIROPTERA		
Emballonuridae		
Sharp-nosed Bat		
Rhynchonycteris naso	B	–
Greater Two-lined Bat		
Saccopteryx bilineata	B	–
Lesser Two-lined Bat		
Saccopteryx leptura	B	B
Peropteryx macrotis	B	B
Diclidurus albus	B	–
Noctilionidae		
Noctilio leporinus	B	B
Mormoopidae		
Pteronotus parnellii	B	B
Pteronotus davyi	B	–
Pteronotus personatus	B	–
Mormoops megalophylla	B	–

ORDER/Family/*Species*	**Trinidad**	**Tobago**
Phyllostomidae		
Micronycteris brachyotis	B	–
Micronycteris hirsuta	B	–
Micronycteris megalotis	B	B
Micronycteris minuta	B	–
Micronycteris nicefori	B	–
Micronycteris sylvestris	B	–
Lonchorhina aurita	B	–
Tonatia bidens	B	–
Tonatia minuta (= *brasiliense?*)	B	–
Mimon crenulatum	B	–
Phyllostomus discolor	B	v
Phyllostomus hastatus	B	B
Trachops cirrhosus	B	–
Vampyrum spectrum	B	–
Glossophaga longirostris	B	B
Glossophaga soricina	B	–
Anoura geoffroyi	B	–
Choeroniscus intermedius	B	–
Carollia perspicillata	B	B
Sturnira lilium	B	B
Sturnira tildae	B	–
Uroderma bilobatum	B	–
Vampyrops helleri	B	–
Vampyrodes caraccioli	B	B
Chiroderma villosum	B	B
Chiroderma trinitatum	B	–
Artibeus cinereus	B	B
Artibeus hartii	B	–
Artibeus jamaicensis	B	B
Artibeus lituratus	B	B
Ametrida centurio	B	–
Centurio senex	B	B
Desmodus rotundus	B	–
Diaemus youngi	B	–
Natalidae		
Natalus tumidirostris	B	B
Furipteridae		
Furipterus horrens	B	–
Thyropteridae		
Thyroptera tricolor	B	–
Vespertilionidae		
Myotis keaysi	B	–
Myotis riparius	B	–
Myotis nigricans	B	B
Eptesicus brasiliensis	B	B
Rhogeessa tumida	B	–
Lasiurus borealis	B	B
Lasiurus ega	B	–

ORDER/Family/*Species*	Trinidad	Tobago
Molossidae		
Molossops greenhalli	B	–
Tadarida europs	B	–
Eumops auripendulus	B	–
Molossus ater	B	–
Molossus major	B	–
Molossus trinitatus	B	–
Molossus molossus	?	B
Promops centralis	B	–
Promops nasutus	B	–
PRIMATES		
Cebidae		
Weeper Capuchin		
Cebus olivaceus	B	–
Red Howler Monkey		
Alouatta seniculus	B	–
CARNIVORA		
Procyonidae		
Crab-eating Raccoon		
Procyon cancrivorus	B	B
Mustelidae		
Tayra		
Eira barbara	B	–
Freshwater Otter		
Lutra longicaudis	B	–
Viverridae		
Small Indian Mongoose		
Herpestes auropunctatus	I	–
Felidae		
Ocelot		
Felis pardalis	B	–
SIRENIA		
Trichechidae		
Caribbean Manatee		
Trichechus manatus	B	–
ARTIODACTYLA		
Tayassuidae		
Collared Peccary		
Tayassu tajacu	B	B
Cervidae		
White-tailed Deer		
Odocoileus virginianus	–	X
Red Brocket		
Mazama americana	B	X
RODENTIA		
Sciuridae		
Neotropical Red Squirrel		
Sciurus granatensis	B	B

ORDER/Family/*Species*	**Trinidad**	**Tobago**
Heteromyidae		
Spiny Pocket Mouse		
Heteromys anomalus	B	B
Muridae		
Arboreal Rice Rat		
Oryzomys concolor	B	B
Terrestrial Rice Rat		
Oryzomys capito	B	–
Pygmy Rice Rat		
Oryzomys delicatus	B	–
Water Rat		
Nectomys squamipes	B	B
Climbing Rat		
Rhipidomys sclateri	B	–
Grass Mouse		
Akodon urichi	B	B
Cane Mouse		
Zygodontomys brevicauda	B	–
Black Rat		
Rattus rattus	I	–
Brown Rat		
Rattus norvegicus	I	–
House Mouse		
Mus musculus	I	–
Erethizontidae		
Prehensile-tailed Porcupine		
Coendu prehensilis	B	–
Dasyproctidae		
Agouti		
Dasyprocta leporina	B	–
Agoutidae		
Paca		
Agouti paca	B	–
Echimyidae		
Trinidadian Spiny Rat		
Proechimys trinitatus	B	–
Arboreal Spiny Rat		
Echimys armatus	B	–

AMPHIBIANS AND REPTILES OF TRINIDAD AND TOBAGO

After Murphy, J. C. 1997. *Amphibians and reptiles of Trinidad and Tobago.* Krieger Publishing Company, Malabar, FL. 245 pp.

AMPHIBIANS

ORDER/Family/*Species*
ANURA
Bufonidae
Bufo beebei Rivero's Toad
Bufo marinus Marine Toad, Cane Toad, Giant Toad, Agua Toad
Centrolenidae
Hyalinobatrachium orientale tobagoensis Tobago Glass Frog
Dendrobatidae
Mannophryne olmonae Bloody Bay Poison Frog
Mannophryne trinitatis Trinidad Poison Frog
Hylidae
Flectonotus fitzgeraldi Mount Tucutche Tree Frog
Hyla boans Gladiator Tree Frog, Spotted Tree Frog, Fasciated Frog
Hyla crepitans Rattle-voiced Tree Frog
Hyla geographica Map Tree Frog
Hyla microcephala misera Yellow Cricket Tree Frog, Small-headed Tree Frog, Underwood's Yellow Tree Frog, Yellow Tree Frog
Hyla minuscula Rivero's Tiny Tree Frog
Hyla minuta Pygmy Tree Frog, Lesser Tree Frog
Hyla punctata Spotted Tree Frog, Polka-dot Tree Frog
Phrynohyas venulosa Veined Tree Frog, Laurenti's Brown Tree Frog, Lost Tree Frog, Milky Tree Frog, Marbled Tree Frog, Marty Tree Frog, Merian Frog
Phyllodytes auratus Golden Tree Frog, Trinidad Heart-tongued Frog
Phyllomedusa trinitatis Trinidad Leaf Frog
Scinax rubra Red-snouted Tree Frog, Allen's Snouted Tree Frog
Sphaenorhynchus lacteus Orinoco Lime Tree Frog, Greater Hatchet-faced Tree Frog
Leptodactylidae
Adenomera hylaedactyla Napo Tropical Bullfrog, Common Chirping Frog
Eleutherodactylus charlottevillensis
Eleutherodactylus johnstonei Johnstone's Robber Frog, Caribbean
Whistling Tree Frog
Eleutherodactylus cf. rozei
Eleutherodactylus urichi Lesser Antilles Robber Frog
Leptodactylus bolivianus Bolivian White-lipped Frog
Leptodactylus fuscus Rufous Frog
Leptodactylus knudseni Knudsen's Frog, Rose-sided Jungle Frog
Leptodactylus macrosternum Miranda's White-lipped Frog
Leptodactylus nesiotus

Leptodactylus validus
Lithodytes lineatus Gold-striped Frog, Chestnut Frog, Painted Antnest Frog
Physalaemus pustulosus Tungara Frog
Microhylidae
Elachistocleis ovalis Common Oval Frog, Oval Frog
Elachistocleis surinamensis Surinam Oval Frog
Pipidae
Pipa pipa Suriname Toad
Pseudidae
Pseudis paradoxa caribensis Caribbean Paradox Frog
Ranidae
Rana palmipes Neotropical Green Frog
GYMNOPHIONA
Typhlonectidae
Typhlonectes sp.

REPTILES

ORDER/Family/*Species*
CHELONIA
Chelidae
Chelus fimbriatus Matamata
Phrynops gibbus Gibba Turtle
Cheloniidae
Caretta caretta Loggerhead Sea Turtle
Chelonia mydas Green Sea Turtle
Eretmochelys imbricata imbricata Hawksbill Sea Turtle
Lepidochelys olivacea Olive Ridley Sea Turtle
Dermochelyidae
Dermochelys coriacea Leatherback Sea Turtle
Emydidae
Rhinoclemmys punctularia punctularia Spot-legged Turtle
Kinosternidae
Kinosternon scorpioides scorpioides Scorpion Mud Turtle
Pelomedusidae
Podocnemis expensa Tartaruga, Arrau Turtle, South American River Turtle, True Amazon Turtle
Podocnemis unifilis Yellow-spotted Amazon River Turtle
Testudinidae
Geochelone carbonaria Red-footed Tortoise
Geochelone denticulata South American Yellow-footed Tortoise
CROCODYLIA
Alligatoridae
Caiman crocodilus crocodilus Spectacled Caiman
Crocodylidae
Crocodylus acutus American Crocodile
Crocodylus intermedius Orinoco Crocodile

SQUAMATA [suborder Amphisbaenia]
Amphisbaenidae
Amphisbaena alba
Amphishaena fuliginous fuliginous
SQUAMATA [suborder Sauria]
Gekkonidae
Gonatodes albogularis Yellow-headed Gecko
Gonatodes cecilae
Gonatodes humeralis Amazon Streak Lizard
Gonatodes ocellatus Eyespot Gecko, Ocellated Gecko
Gonatodes vittatus vittatus Striped Leaf Gecko
Gymnodactylus geckoides
Hemidactylus mabouia Amerafrican House Gecko, Afroamerican House Gecko, Afro-american House Gecko, Tropical House Gecko, Cosmopolitan House Gecko, Moreau's Tropical House Gecko, Phantom Gecko
Hemidactylus palaichthus Antilles Leaf-toed Gecko
Sphaerodactylus molei Mole's Gecko, Tobago Least Gecko
Thecadactylus rapicauda Turnip-tailed Gecko
Gymnophthalmidae
Bachia flavescens
Bachia heteropa alleni Allen's Ground Lizard
Bachia heteropa trinitatis
Gymnophthalamus speciosus Golden Spectacled Lizard
Gymnophthalamus underwoodi
Proctoporus shrevei Luminous Lizard
Iguanidae
Anolis aeneus Smooth-scaled Worm Lizard
Anolis chrysolepis planices
Anolis extremus Extreme Anole, Barbados Anole
Anolis cf. lemurinus
Anolis richardii Richard's Anole
Anolis trinitatis Trinidad Anole
Iguana iguana Green Iguana
Polychrus marmoratus Many-colored Bush Anole
Tropidurus plica Collared Tree Lizard, Collared Tree Runner
Scincidae
Mabuya bistriata Two-lined Skink
Teiidae
Ameiva ameiva Giant Ameiva
Cnemidophorus lemniscatus Rainbow Whiptail
Kentropyx striatus
Tupinambis teguixin Common Tegu, Colombian Tegu, Black Tegu
SQUAMATA [suborder Ophidia]
Anomalepididae
Helminthophis sp. Yellow-headed Ground Puppy
Boidae
Boa constrictor Constrictor Mantona, Redtail Boa, True Redtail Boa, Brazilian Redtail, Surinam Redtail, Peruvian Redtail, Macajuel, Boa Constrictor

Corallus hortulanus cookii Amazon Treeboa
Corallus ruschenbergerii Cascabel Dormillon, Tree Boa
Epicrates cenchria maurus Columbian Rainbow Boa, Rainbow Boa
Eunectus murinus gigas Green Anaconda, Huille, Anaconda

Colubridae

Atractus trilineatus Three-lined Ground Snake
Atractus univittatus One-banded Ground Snake
Chironius carinatus carinatus Lined Keeled Racer, Machete Savane
Chironius multiventris septentrionalis Long-tailed Machete Savane
Clelia clelia clelia Mussurana, Black Cribo
Dipsas variegata trinitatis Trinidad Thirst Snake, Snail-eating Snake
Drymarchon corais corais Eastern Indigo Snake, Yellow-tail Cribo
Erythrolamprus aesculapii aesculapii Common False Coral Snake
Erythrolamprus bizona False Coral Snake
Erythrolamprus ocellatus False Coral Snake, Tobago False Coral, Red Snake
Helicops angulatus Banded Neotropical Water Snake, Banded South American Water Snake, Water Mapepire
Hydrops triangularis neglectus Common Mudsnake, Water Coral
Imantodes cenchoa cenchoa Blunthead Tree Snake, Bluntheaded Tree Snake, Mapepire Corde Violon, Fiddle-string Snake
Leptodeira annulata ashmeadi Common Cat-eyed Snake, False Mapepire
Leptophis ahaetulla coeruleodorsus Black-skinned Parrot Snake, Lora, Parrot Snake
Leptophis riveti Turquoise Parrot Snake
Leptophis stimsoni Gray Lora
Liophis cobella cobella (cobellus cobellus) Banded Tricolored Snake, Mangrove Snake, Mapepire Mangue
Liophis melanotus nesos Beh Belle Chemin, Doctor Snake (Tobago)
Liophis reginae ssp. Common Swampsnake
Liophis reginae zweifeli Common Swampsnake, High Woods Coral
Mastigodryas boddaerti boddaerti Tan Racer, Machete Couesse
Mastigodryas boddaerti dunni Tan Racer, Machete Couesse
Ninia atrata Red-nape Snake, Ring Neck Snake
Oxybelis aeneus Mexican Vine Snake, Horsewhip
Oxyrhopus petola petola Calico Snake, False Coral
Pseudoboa neuwiedii Coal Snake, Neuwied's False Boa, Ratonel
Pseustes poecilonotus polylepis Common Birdsnake, Dos Cocorite
Pseustes sulphureus sulphureus Amazon Puffing Snake, Yellow-bellied Puffing Snake
Sibon nebulata nebulata Cloudy Snail-eating Snake, Clouded Snake
Siphlophis cervinus Common Liana Snake, Checkerbelly
Spilotes pullatus pullatus Common Tiger Ratsnake, Tigre, Tigro
Tantilla melanocephala Common Black-headed Snake
Thamnodynastes sp. Striped Swamp Snake
Tripanurgos compressus Red-eyed Treesnake, Mapepire De Fe

Elapidae

Micrurus circinalis Common Coralsnake, Common Coral Snake
Micrurus lemniscatus diutius South American Coralsnake, Large Coral Snake

Leptotyphlopidae
Leptotyphlops albifrons White-fronted Threadsnake, Ground Puppy
Typhlopidae
Typhlops brongersmianus Black Blindsnake, Burrowing Snake
Typhlops trinitatus Trinidad Blindsnake, Trinidad Burrowing Snake
Viperidae
Bothrops asper Common Lancehead, Fer-de-lance
Lachesis muta muta Amazon Bushmaster, Mapepire Zanana

SELECTED BIBLIOGRAPHY

Birds

Barrow, R. 1994. *Birds of Trinidad & Tobago; A Photographic Atlas.* Media & Editorial Projects Ltd., Maraval, Trinidad. ISBN 9768050098 (hardcover), 976805008X (paperback). A rather large 'coffee table' book; a remarkable combination of exquisite photographs and lucid descriptions of some of the more colourful bird species of Trinidad and Tobago.

Dinsmore, J. J. 1972. *Avifauna of Little Tobago Island. Quart. J. Fla. Acad. Sci.* 35(1):55-71. What can an ornithologist see in several months spent alone on a tiny tropical island? Dinsmore herein describes such an experience.

ffrench, R. 1991. *A Guide to the Birds of Trinidad & Tobago.* 2nd edition. Cornell University Press, Ithaca, New York. ISBN 0801497922. Essential for visiting birders. More than a field guide, it contains background information about Trinidad and its ecology as well as a great deal of detailed information on the various bird species.

ffrench, R. A. 1986. *Birds of Trinidad and Tobago.* Caribbean Publishing, Grand Cayman, Cayman Islands. ISBN 0333409124. A short, lightweight guide to the birds of Trinidad and Tobago for beginning birders. The birds featured in the book are arranged according to the type of habitat they are found in (gardens, forests, etc.) which makes it very easy to use.

Hilty, S. L. 1994. *Birds of Tropical America: A Watcher's Introduction to Behavior, Breeding and Diversity***.** Chapters Publishing Ltd., Shelburne, Vermont. ISBN 1881527565. A wide-ranging exploration of some of the most interesting sites in the Neotropics. Intriguing investigations of fascinating bird behavior. Written by one of the Neotropics' foremost students of ornithology.

Hilty, S.L. 2003. *Birds of Venezuela.* Christopher Helm, London. ISBN 0-7136-64118-5. A superb new field guide to the bird of adjacent Venezuela. All most all of Trinidad's birds are well covered here, with excellent plates and a fine text, the sections on identification, voice, and behaviour all being of great relevance to birders visiting Trinidad and Tobago.

Snow, D.W. 1976. *The Web of Adaptation: Bird Studies in the American Tropics.* Collins, London. ISBN 0002197359. If you only read one book before your visit, read this one. It is composed of many short chapters, each one covering a particular facet of tropical birds.

Snow, D.W. 1982. *The Cotingas.* Oxford University Press, Oxford. ISBN 019858511X. An examination of the unusual assortment of species, including the bellbirds, that are lumped in the avian order Cotingidae. While writing this book David Snow and his wife lived about 5 km down the valley from the Asa Wright Nature Centre.

Mammals

Eisenberg, J. F. 1989. *Mammals of the Neotropics: Volume 1 - The Northern Neotropics.* University of Chicago Press, Chicago. ISBN 0-226-19539-2 (hardback), ISBN 0-226-19540-6 (paperback). A coffee-table-sized book to the mammals of northern South America including Trinidad and Tobago. Contains good colour plates and covers some species not included in the following title.

Emmons, L. H. 1990. *Neotropical Rainforest Mammals.* University of Chicago Press, Chicago. ISBN 022620718-8. Excellent field guide to the mammals that occur in rainforests throughout South America.

Amphibians and Reptiles

Boos, H. E. A. 2001. *The Snakes of Trinidad and Tobago.* Texas A & M University Press, College Station, Texas. ISBN 1585441163. Hans Boos managed the Emperor Valley Zoo in Port-of-Spain and assisted with expeditions studying birds, reptiles, and amphibians. In this book he covers more than 60 species of snakes found in Trinidad and Tobago. The culmination of thirty years studying the snakes of these islands, this book provides in-depth information.

Murphy, J. C. 1997. *Amphibians and Reptiles of Trinidad & Tobago.* Krieger Publishing Company, Melbourne, Florida. ISBN 089464971X. Covers the entire herpetofauna of Trinidad and Tobago describing 130 species and subspecies and their natural history. Includes descriptions and distributions with maps, a section of general information describing the environments of the islands, 170 colour photos, and identification keys.

Reed, M. 2004. Frogs of Trinidad. A web-site with colour photographs of many of Trinidad's frogs.
http://uk.geocities.com/craspedopus/frogs_on_line/trinfrogs.html

Reed, M. 2004. Frogs of Tobago. A web-site with colour photographs of many of Tobago's frogs.
http://uk.geocities.com/craspedopus/frogs_on_line/tobagofrogs.html

Insects

Barcant, M. 1970. *Butterflies of Trinidad & Tobago.* Collins, London. ISBN 0-00-212027-5. The standard field guide, covering hundreds of species but sadly long out-of-print. Second-hand copies can occasionally be bought from specialist book dealers.

Fish and Sea-life

Chaplin, C. C. G. 1972. *Fishwatcher's Guide to West Atlantic Coral Reefs.* Harrowood Books, Valley Forge, Pennsylvania. ISBN 0915180081. Printed on waterproof material. Out-of-print.

Colin, P. I. and P. L. Colin. 1978. *Caribbean Reef Invertebrates and Plants: A Field Guide to the Invertebrates and Plants Occurring on Coral Reefs of the Caribbean, the Bahamas and Florida.* T. F. H. Publications, Neptune City, New Jersey. ASIN 0876664605. Good source of information for those preparing to snorkel or scuba dive around Trinidad and Tobago.

Greenberg, I. and J. Greenberg. 1977. *Guide to Coral and Fishes of Florida, the Bahamas and the Caribbean.* Seahawk Press, Miami, Florida. ASIN 0913008087 (paper), ASIN 0913008079 (waterproof). A fairly comprehensive volume providing identification for almost all the underwater life around Trinidad and Tobago.

Greenberg, I. and J. Greenberg. 1981. *Sharks and Other Dangerous Sea Creatures.* ISBN 0913008095. Exciting, full-colour photos and illustrations and accurate text give the real lowdown on the most dangerous denizens of the deep – sharks, eels, barracudas, and downright poisonous creatures – many as beautiful as they are deadly. Covers shark attack and research plus most of the potentially hazardous sea life found in the waters of Florida, the Bahamas, and the Caribbean.

Greenberg, I. et al. 1987. *The Fishes Beneath Tropic Seas.* ISBN 0913008184. Seahawk Press, Miami, Florida. A unique blend of artwork by Idaz Greenberg with photographs by Jerry and Michael Greenberg portray the vivid splendour of the reef. Scientifically accurate text makes this a must-have for divers and marine naturalists. Covers the waters of Florida, the Bahamas, and the Caribbean.

Humann, P. and N. DeLoach. 2001. *Reef Coral Identification: Florida, Caribbean, Bahamas.* 2nd edition. New World Publications, Inc. ISBN 1878348329. This book helps identify virtually every species of stony coral, gorgonian, fire coral, and black coral inhabiting the tropical western Atlantic.

Humann, P. and N. DeLoach. 2001. *Reef Creature Identification: Florida, Caribbean, Bahamas.* 2nd edition. New World Publications Inc., Jacksonville, Florida. ISBN 1878348310. Identifies the weird and wonderful crabs, lobsters, shrimps, jellies, sponges, tunicates, and molluscs that inhabit the reefs of the Caribbean.

Humann, P. and N. DeLoach. 2002. *Reef Fish Identification: Florida, Caribbean, Bahamas.* 3rd edition. New World Publications Inc. ISBN 1878348302. According to many scuba-divers, the best book on this subject available today.

Kenny, J. S. 1995. *Views from the Bridge: A Memoir on the Freshwater Fishes of Trinidad.* Published by the author, Maracas, Trinidad. ISBN 976-8136-74-X. An account of the distribution and ecology of Trinidad freshwater fish, including distributional maps, the author's view on the historical basis for the distributional patterns, species accounts, photographs of 36 of the 48 freshwater species, and an account of attempts to farm fish in Trinidad. Contains many anecdotes about people who have worked on fish in Trinidad, their collection sites, and study areas. A valuable reference for anyone interested in Neotropical and/or Trinidadian fish.

Laydoo, R. S. 1991. *A Guide to the Coral Reefs of Tobago.* Institute of Marine Affairs and the Asa Wright Nature Centre, Arima, Trinidad. ASIN 9768104015.

Lieske, E. and R. Myers. 2001. *Coral Reef Fishes: Indo-Pacific and Caribbean.* Princeton University Press, Princeton, New Jersey. ISBN 0691089957. A handy guide to all fishes likely to be observed by anyone visiting or diving on the coral reefs of the Caribbean.

Martini, F. 1984. *Exploring Tropical Isles and Seas: An Introduction for the Traveler and Amateur Naturalist.* Prentice-Hall, Englewood Cliffs, New Jersey. ISBN 0132959313. An introduction to tropical coral reefs, tropical islands, and tropical oceans, with their respective marine mammals, reptiles, and fishes.

Ramjohn, D. D. 1999. *Checklist of Coastal and Marine Fishes of Trinidad & Tobago.* Fisheries Information Series No. 8, Fisheries Division, Ministry of Agriculture, Land and Marine Resources, St. Clair Circle, Port-of-Spain. A reference document for all species of fish identified from the waters of Trinidad and Tobago.

Robins, C. R. and G. C. Ray. 1986. *A Field Guide to Atlantic Coast Fishes of North America (Peterson Field Guide Series, 32).* Houghton Mifflin Co., Boston. ISBN 0395975158. More than 1,000 species are described in this guide including information on range and habitat such as depths, bottom types, water temperatures, and salinity. The 1,100 illustrations use the Peterson Identification System for quick, accurate field identification. Almost all the marine gamefish in North America can be found off Trinidad and Tobago, so this book may help to identifying them.

Sutty, L. 2000. *Seashell Treasures of the Caribbean.* Macmillan Caribbean. ISBN 0333745280. Thorough coverage of a great many beautiful species.

Voss, G. L. (ed.). 2002. *Seashore Life of Florida and the Caribbean.* Dover Publications, Inc. ISBN 048642068X. Includes coverage of Trinidad and Tobago.

Wood, L. 2000. *Diving & Snorkelling Trinidad & Tobago.* Lonely Planet, Oakland, California. ISBN 0864427778.

Flora

Beard, J. S. 1944. *The natural vegetation of the island of Tobago, British West Indies. Ecol. Monographs 14:135-163.* A thorough treatment of the four types of forests found on Tobago as well as descriptions of the vegetation in bygone eras.

Bourne, M. J. et al. 1988. *Fruits and Vegetables of the Caribbean.* MacMillan Publishers Ltd., London. ISBN 0333453115.

Dennis, R. W. G. 1970. *Fungus Flora of Venezuela and Adjacent Countries.* William Clowes and Sons, Ltd., London. ASIN 0112410006.

Duncan, E. J. 1993. *A Guide to the Wild Flowers of Trinidad & Tobago.* Asa Wright Nature Centre, Arima, Trinidad. A splendid photographic guide to some of the most conspicuous wildflowers of these islands. Arranged by flower colour.

Hargreaves, D. and B. Hargreaves. 1960. *Tropical Blossoms of the Caribbean.* Hargreaves Company, Inc., Kailua, Hawaii. ASIN 0910690030. Out-of-print.

Hargreaves, D. and B. Hargreaves. 1965. *Tropical Trees Found in the Caribbean, South America, Mexico.* Hargreaves Company, Inc., Kailua, Hawaii. ISBN 0910690057. Out-of-print but occasionally available through used-book dealers.

Kenny, J. S. 1988. *Native Orchids of the Eastern Caribbean (Caribbean Pocket Natural History Series).* MacMillan Caribbean. ISBN 0333473302. Over 100 orchids are described and beautifully illustrated in this book, all of which were photographed in Trindad and Tobago, despite the title.

Lotschert, W. and G. Beese. 1999. *Collins Guide to Tropical Plants: A Descriptive Guide to 323 Ornamental and Economic Plants with 274 Colour Photographs.* HarperCollins, London. ASIN 0002191121. A comprehensive identification guide to all major orders of tropical plants.

Mahabir, N. K. 2001. *Medicinal and Edible Plants Used by East Indians of Trinidad & Tobago.* Chakra Publishing House, San Juan, Trinidad. ISBN 9768001739. Information on the medicinal use 65 local plants, each of which is described and given a botanical name.

Seddon, S. A. and G. W. Lennox. 1980. *Trees of the Caribbean.* Macmillan Education Ltd., London. ISBN 0333287932. A photographic guide to some of the more easily identified and widespread trees of the Caribbean.

Natural History/Ecology

Beebe, W. 1952. *Introduction to the ecology of the Arima Valley, Trinidad. Zoologica 37:157-183.* A highly readable description of the area of the Asa Wright Nature Centre, written by the famous North American naturalist.

Caufield, C. 1986. *In the Rain Forest: Report from a Strange, Beautiful, Imperiled World.* University of Chicago Press, Chicago. ISBN 0226097862. A comprehensive account of the status of the world's richest ecological communities, past, present, and future.

Cooper, G. C. and P. R. Bacon. 1981. *The Natural Resources of Trinidad and Tobago.* Edward Arnold, London. Outlines the country's physical, biological, and human resources and includes useful appendices.

Forsyth, A. and K. Miyata. 1984. *Tropical Nature.* MacMillan Publishing Company, New York. ISBN 0684187108. Fascinating introduction to the plant and animal life of the wet tropics. Especially valuable to all interested in the study of this lush habitat.

Kricher, J. 1989. *A Neotropical Companion: An Introduction to the Animals, Plants, and Ecosystems of the New World Tropics.* Princeton University Press, Princeton, New Jersey. ISBN 0691044333. An introduction to the animals, plants, and ecosystems of the New World tropics. Broad coverage, fun reading.

Worth, C. B. 1967. *A Naturalist in Trinidad.* J. B. Lippincott, Philadelphia. Out-of-print but sometimes available in used book stores. A wonderful book composed of short chapters with titles like Monkeys, Small Mammals, Bird Netting, the Arima Valley, and Hummingbirds.

Travel guides

de-Light, D. and P. Thomas. 1998. *The Rough Guide to Trinidad and Tobago.* A general-purpose travel guide. Rough Guides, New York.

Doyle, C. 2001. *Cruising Guide to Trinidad and Tobago.* 3rd edition. Cruising Guide Publications, Dunedin, Florida. ISBN 0944428592. As Trinidad and Tobago become better known as a place for yachting enthusiasts, a book like this becomes more and more valuable. Provides all the information a yachtsman could need, with colour photos of some of Trinidad and Tobago's most beautiful bays.

Insight Guides. 2000. *Trinidad & Tobago.* 3rd edition. APA Publications, Singapore. ISBN 0887298087. A superbly colourful guide to the history and culture of Trinidad & Tobago.

O'Donnell, K. and S. Pefkaros. 2000. *Adventure Guides to Trinidad & Tobago.* Hunter Publishing, Inc., Boston, Massachusetts. ISBN 1556508867. Most Trinidad and Tobago guides assume that all visitors want to do is sunbathe so only provide lists of hotels, directions to beaches, and scant mention of bird sanctuaries. These authors explore the adventurous side of the islands as well, with hiking and kayaking, mud volcanoes, waterfalls, and swamps.

Pavlidis, S. J. 2002. *A Cruising Guide to Trinidad and Tobago.* Seaworthy Publications, Port Washington, Wisconsin. ISBN 189239913X. Similar in most respects to C. Doyle's *Cruising Guide to Trinidad and Tobago* (see above).

Language

Baptiste, R. 1993. *Trini Talk: A Dictionary of Words and Proverbs of Trinidad and Tobago.* Caribbean Information Systems & Services, Port-of-Spain. ISBN 9768133015. A dictionary of words and proverbs with excellent examples.

Thomas, J. J. 1969. *The Theory and Practice of Creole Grammar.* New Beacon Books, London. ISBN 0901241040. Interest in the 'Nation Language' of Trinidad and Tobago has elicited widespread academic interest during the last few decades. The author, a son of freed slaves, was proficient in several languages.

History and culture

Anthony, M. 1997. *Historical Dictionary of Trinidad & Tobago.* Scarecrow Press, Lanham, Maryland. ISBN 0810831732. Describes the important people, places, events, and institutions of the nation as well as its society, culture, and economy. Includes maps and a chronology content that is both current and retrospective.

Anthony, M. 1988. *Towns and Villages of Trinidad and Tobago.* Circle Press, Port-of-Spain. Spanning the historical as well as the present, this book examines approximately 60 towns and villages and includes maps of some of the more important ones.

Besson, G. 1992. *The Book of Trinidad.* Paria Publishing, Port-of-Spain. ASIN 9768054360. Covers all the ethnic groups in Trinidad – African, Amerindian, East Indian, and European. Chronicles Trinidad's history, examines its culture, economy, and sociology, and profiles its most influential individuals.

Campbell, C. 1992. *Colony and Nation: A Short History of Education in Trinidad and Tobago.* Ian Randle, Kingston, Jamaica. ISBN 9768100052. Traces the history of education in Trinidad and Tobago from slave society to the mid-1980s, with special focus on women and education under Eric Williams.

De Boissierre, R. 1982. *Crown Jewel.* Allison & Busby, London. ISBN 0805280995. Moving narrative of a charismatic black activist and his attempts to unify blacks in Trinidad for social progress.

Forte, M. 1997. *Against the Trinity: An Insurgent Imam Tells His Story: Religion, Politics and Rebellion in Trinidad & Tobago.* Ahead Desktop Publishing House, Binghampton, New York. ISBN 1888024119.

Hill, E. 1997. *The Trinidad Carnival.* New Beacon Books, London. ISBN 1873201141. The first comprehensive work published on Carnival in all of its aspects, placing Carnival in the context of the development of the Trinidad and Tobago persona.

Lovelace, E. 1998. *Salt.* Persea Books, New York. ISBN 0892552352. This short story writer, novelist, and playwright from Trinidad and Tobago has emerged as one of the region's great novelists. This book has been described as 'a triumphant, vibrant novel peopled by enduring and embattled women and men striving with passion and wit to make sense of the intimacies and the contradictions of their lives in the still young country of Trinidad, where the roles of enslaved and landowner linger, yet where the sky, the sea, every green leaf and tangle of vine sing freedom'.

Mavrogordato, O. J. 1996. *Voices in the Street*. Paria Publishing Co., Port-of-Spain. ASIN 976805445X. First published in 1977, this edition is lavishly illustrated with photographs from the well-respected Norton Studio. Copious information about historic sites around Queen's Park Savannah, especially the Magnificent Seven. Provides a very brief account of the history of Trinidad and Tobago explored largely through places and memoirs of Trinidad's upper classes.

Naipaul, V. S. 2001. *A House for Mr. Biswas*. Vintage Books, New York. ISBN 0375707166. A classic work that provides an intimate and accurate look at a the life of a typical Hindu Trinidadian half a century ago.

Naipaul, V. S. 1993. *The Middle Passage: Impressions of Five Societies – British, French, and Dutch – In the West Indies and South America*. Penguin USA (paper). ISBN 0140029206. A classic of modern travel writing. Naipaul creates an incredibly accurate portrait of Trinidad and four adjacent Caribbean societies that still remember slavery and colonialism and are so strongly tied to the traditions of Empire that they are in denial that the days of Empire are numbered.

Richardson, B. C. 1992. *The Caribbean in the Wider World, 1492-1992: A Regional Geography (Geography of the World Economy)*. Cambridge University Press, Cambridge, Massachusetts. ISBN 0521351863. Explains how the region's present geography is intimately tied to the past. The Caribbean was Europe's first overseas colony, its landscapes transformed to produce tropical staples and its decimated aboriginal populace replaced with African slaves. As European power has waned in the Caribbean it has been replaced by the geopolitical domination of the United States.

Rogozinski, J. 1994. *A Brief History of the Caribbean; From the Arawak and the Carib to the Present*. Meridian Books, New York. ISBN 0452011345. Provides a descriptive and analytical history of the Caribbean islands, focusing on the interrelated evolution of the economy, political institutions, and social forms from the landing of Columbus in the Bahamas 500 years ago to the present.

Rudder, J. 1991. *Our Native Land*. United Nations Environment Programme, Nairobi. A detailed study of global environmental concerns such as exploitation of wildlife, water pollution, and environmental accountability and their relevance to Trinidad and Tobago.

Samaroo, B. and K. Ramchand (eds.). 1995. *In Celebration of 150 Years of the Indian Contribution to Trinidad and Tobago*. Estorical Publications Ltd., Port-of-Spain. (Includes the *Indian Centenary Review*, which examined the Indian in Trinidad and Tobago from 1845 to 1945.)

Includes information on Indians and Indian culture, place and personal names and their meanings, as well as historical information.

Selvon, S. 1995. *Brighter Sun.* Dearborn Financial Publishing Inc., New York. ISBN 0582642655. The master of the Trinidad and Tobago 'Nation Language', Selvon's characterisations are considered to be quite beautiful. This book describes the struggles of a young Hindu man trying to forge a life on his own with his young wife in the 1940s. His interactions with the African couple next door provide insights into different cultures and the possibilities of racial confusion.

Toth, C. W. 1975. *The American Revolution and the West Indies.* Kennikat Press, Port Washington, New York. ISBN 080469110X. The sugar-molasses-rum triangle and its profound effect on the history of North America.

Warner-Lewis, M. et al. 1991. *Guinea's Other Suns: The African Dynamic in Trinidad Culture.* Majority Press, Dover, Massachusetts. ISBN 0912469277. Documents the Africans in Trinidad and Tobago society with focus on the Orisa religion and the Yoruba in Trinidad and Tobago. It examines Yoruba chants and their translation into English and the African elements in Trinidad and Tobago's 'Nation Language'.

Williams, E. 1994. *Capitalism and Slavery.* University of North Carolina Press, Raleigh. ISBN 0807844888. The acknowledged father of the republic of Trinidad and Tobago was also a towering intellectual. This seminal work is part of the necessary rewriting of Caribbean history by Caribbean scholars. Provides an economic study of the role of African slavery and the slave trade. This book is credited with debunking the myth that emancipation was solely the result of the efforts of the anti-slavery movement and a quickening of conscience, but rooted instead in the economic reality of the time.

Williams, E. 1984. *From Columbus to Castro: The History of the Caribbean 1492 to 1969.* Vintage Books, New York. ISBN 0394715020. Interesting and well-written essay on slavery and sugarcane and the effects of their interaction on the peoples of the Caribbean.

Williams, E. 1993. *History of the People of Trinidad & Tobago.* A & B Book Pub. Dist. ISBN 1881316653. A classic study, written by a scholar who from 1961 until his death in 1981 served as Prime Minister of Trinidad and Tobago.

Woodcock, H. I. 1971. *A History of Tobago.* Frank Cass & Co. Ltd., London. ISBN 071461954X. First published in 1867, this history of Tobago was written by a man who came to Tobago as Chief Justice of the island.

LOCAL CONTACTS AND SOCIETIES

Local Guides

A number of local people make a living working as birdwatching guides on Trinidad and Tobago. In addition to knowing the birds, these guides can take visitors to specific sites for hard-to-find species and congregation sites ('leks' and roosts) and can point out special behaviours of interesting species. In Trinidad, and even more so in Tobago, being seen in the company of a local guide can warm the local people towards visitors, especially in remote areas where outsiders are few.

Jogie Ramlal

For years Jogie Ramlal, Rudal Ramlal, and Winston Nanan have served as the native guides of choice on Trinidad. Jogie is an excellent all-around landbirding guide, having lived for many years in Verdant Vale, just down the Arima Valley from the Asa Wright Nature Centre. He has led thousands of field trips to all parts of Trinidad and possesses extensive knowledge of songs, habits, and haunts of Trinidad's birds. Jogie has discovered several species of birds new to the island. His peaceful, friendly demeanor, remarkable serenity, and fine sense of humour make him a wonderful companion in the field. Jogie is one of several guides currently utilized by the Asa Wright Nature Centre.

Mahase Ramlal

Jogie's son, Mahase, is also an accomplished birder and guide as is Jogie's brother, Rudal Ramlal, both of whom guide for the Asa Wright Nature Centre.

Nanan Tours has been associated with the Caroni Swamp for more than half a century. Winston Nanan is well versed in the natural history of the entire Caroni Swamp ecosystem. Winston has spent his life close to his primary birding area and thus has an unmatched and intimate knowledge of the Caroni Swamp. He is one of the few native naturalists who is equally familiar with the fauna and flora of Tobago and of northern South America.

Winston Nanan

Courtenay Rooks

Courtenay Rooks has spent a lifetime developing an intimate relationship with the natural history of Trinidad and Tobago. He has been leading birding tours in the islands since 1993. He is Chair of the Bird Group, Trinidad and Tobago Field Naturalists' Club, serves on the Trinidad and Tobago Rare Bird Committee. He is also active in conservation and ecotourism, serving as managing director and founder of the Paria Springs Project, which involves the construction of an ecolodge in the middle of Trinidad's Northern Range. Courtenay also developed a project dedicated to the conservation of the Ocelot and of Trinidad's natural habitats for birds.

Several excellent nature guides are associated with the Asa Wright Nature Centre. At the time of writing these include Molly Calderon, Harold Diaz, Denise Etienne, James Maule, Jason Radix, and Mukesh Ramdass.

Roger Neckles is a Trinidadian naturalist best known for his photography. Photographs by Roger can be seen at Piarco Airport in Trinidad and Crown Point Airport in Tobago. Roger formed his own ecotourism company, Avifauna Tours (http://www.trinidad.net/avifauna/) a few years ago.

Purple Honeycreeper. *(Photo by Roger Neckles)*

Adolphus James

Adolphus James began leading birding trips in 1966 and is the Asa Wright Nature Centre's guide for Tobago. His spacious maxi-taxi can accommodate birding groups in real comfort. The windows are curved and continue almost to the roof, allowing good views of the surroundings.

David Rooks

Another nature guide in Tobago is David Rooks. David is a former president of the Trinidad and Tobago Field Naturalists Club, the oldest natural-history club in the Western Hemisphere. David caters for groups of all sizes.

Two other excellent Tobago guides are Peter Cox and Newton George. Both know Tobago intimately and their remarkable enthusiasm and sense of humour makes a day in the field with them a day to be long remembered.

If you would like a guide along Gilpin Trace on the Main Ridge, Frankie Leander and his associates can be found at the site.

Left to right: Peter Cox and Newton George

For those visiting Tobago without the services of an expert bird guide and unable to secure the services of Adolphus James or one of his sons with a maxi-taxi, Peter

Moore is highly recommended as a provider of transportation. Peter will simply drive visitors quickly and safely to the island's top birding spots, where birders can locate and identify the birds themselves.

Frank's Boat Tours are recommended for trips from Speyside to Little Tobago Island. Wordsworth Frank has been guiding naturalists around northeastern Tobago for many years. Frank's Boat Tours also run snorkelling trips.

In Trinidad

For landbirding:

Courtenay Rooks
44 La Seiva Road
Maraval, Trinidad
(868) 622-8826 voice
(868) 628-1525 fax
e-mail: rooks@pariasprings.com
web-site: http://www.pariasprings.com/rookstours

Jogie Ramlal
Mahase Ramlal
Rudal Ramlal
Milepost 3¾, Blanchisseuse Road
Arima, Trinidad
(no telephone; contact through Asa Wright Nature Centre)

Denise Etienne
Wade Hamilton
James Maule
Jason Radix
Mukesh Ramdass
Chima Rudder
Vijay Singh
c/o Asa Wright Nature Centre
P.O. Box 4710
Arima, Trinidad
(868) 667-4655 voice
(868) 667-0493 fax

Roger Neckles
17 Morne Haven Condominiums
Gilkes Street
Morne Coco Road
Diego Martin, Trinidad
(868) 633-5614

For Caroni Swamp boat tours and other nature tours throughout the southern Caribbean:

Winston Nanan Bird Sanctuary Tours
Bamboo Grove Settlement No. 1
Princess Margaret Highway
Curepe P.O., Trinidad
(868) 645-1305 voice and fax
e-mail: nantour@tstt.net.tt
or contact Tropical Birding Adventures Inc.

In Tobago

Peter Cox
Milford Road
Canaan, Tobago
(868) 639-3093 home
(868) 751-5822 cell
e-mail: petercoxislandtours@excite.com

Wordsworth Frank
Frank's Boat Tours
Speyside Main Road
Speyside, Tobago
(868) 660-5438

Newton George
#3 Top Hill Street
Speyside, Tobago
(868) 660-5463
e-mail: ngeorge@tstt.net.tt

Adolphus James
P.O. Box 346
Scarborough Post Office, Tobago
or at Crown Point Airport (license HAL 9826)
(868) 639-2231 voice and fax

Frankie Leander
Apple Trace
Roxborough, Tobago
(868) 660-4426

Peter Moore (expert driver)
192 Mora Avenue
Milford Court
Bon Accord, Tobago
(868) 639-2312 voice
(868) 660-2312 answering machine

David and Courtenay Rooks
44 La Seiva Road
Maraval, Trinidad
(868) 622-8826 voice
(868) 628-1525 fax
e-mail: rooks@pariasprings.com
web-site: http://www.pariasprings.com/rookstours

or

P.O. Box 348, Scarborough, Tobago
(868) 639-4276 voice
(868) 639 5440 fax
e-mail: rookstobago@trinidad.net
web-site: http://www.trinidad.net/rookstobago

Other nature tour companies in Trinidad and Tobago:

Caribbean Discovery Tours (Steven Broadbridge)
(Specialist in environmentally sensitive adventure travel.)
9-B Fondes Amandes Road, St. Ann's, Trinidad
e-mail: info@caribbeandiscoverytours.com
web-site: http://www.caribbeandiscoverytours.com

South East Ecotours (Keith Rodriguez)
(Specializes in nature and environmentally sensitive tours in southeastern Trinidad.)
Fuentes Street, Rio Claro, Trinidad
(868) 644-1072 voice
(868) 644-1072 fax
e-mail: secotour@tstt.net.tt

Wildways (Wendy Yawching, Andy Whitwell)
(Specialist in environmentally sensitive adventure travel.)
10 Idlewild Road, Knights Bridge, Cascade, Trinidad
(868) 623-7332 phone and fax
e-mail: wildways@trinidad.net
web-site: http://www.wildways.org

TRINIDAD AND TOBAGO FIELD NATURALIST CLUB

Although no organization exists in Trinidad and Tobago exclusively for the study of birds, the Trinidad and Tobago Field Naturalists' Club (http://www.wow.net/ttfnc/), founded in 1891, caters to all those interested in natural history. Lectures are held on the second Thursday of each month at 5:30 p.m. in the Audio Visual Room of St. Mary's College, Frederick Street, Port-of-Spain. These meetings are well worth attending. Field excursions to interesting sites in Trinidad and Tobago are conducted on the last Sunday of every month except December, departing from the rear entrance of St. Mary's College, Pembroke

Street, Port-of-Spain, at 6 a.m. For field trip information contact Courtenay Rooks at (868) 622-8826 or e-mail rooks@pariasprings.com. The Club's special-interest groups conduct research projects and publish the results of their investigations. They also publish articles on environmental matters and participate in exhibitions, book fairs, and special events. In addition, the Club provides advisory services to public and private organizations on environmental matters.

Membership is open to anyone 15 years and over who subscribes to the club's objectives. Additional information is obtainable from The Secretary, Trinidad and Tobago Field Naturalists' Club, Box 642, Port-of-Spain, Trinidad.

TRINIDAD AND TOBAGO RARE BIRD COMMITTEE

Records of unusual species should be submitted to the Trinidad and Tobago Rare Bird Committee (TTRBC) (http://www.wow.net/ttfnc/ttrbc/ttrbcHome.html). The TTRBC web site contains information about the committee members, the species review list, the official bird list, tips for taking field notes, and published reports. The site also provides links to the Southeastern Caribbean Bird Alert, to rare bird committees in other countries, and to a wide range of other interesting material and sites.

For documenting rare birds, photographs or video footage is helpful, but comprehensive written notes and sketches are also welcome, especially those made in the field at the time. The Committee requests that the following information be included with your submission: date, time, location (as specifically as possible), name of observer(s), address(es), and details of the sighting, including length of time the bird was observed or heard and distance from it (them), lighting conditions, optical equipment used, and familiarity of the observer(s) with the species and with similar species. Include anything and everything seen and heard. Details that observers note but which *do not appear* in field guides are often familiar to committee members and sometimes enable them to accept borderline submissions.

A printable copy of the Rare Bird Report Form is available on the Internet at http://www.wow.net/ttfnc/ttrbc/ttrbcFrm.html. Please report appropriate sightings promptly to:

Martyn Kenefick
36 Newalloville Avenue
San Juan, Trinidad and Tobago
e-mail: martynkenefick@hotmail.com
fax: (868) 674-7609 voice

A TRINBAGONIAN LEXICON

English is the official language of the Republic of Trinidad and Tobago but many common Trinbagonian words are derived from African, Amerindian, Arabic, East Indian, French, Spanish, and Portuguese. This conglomerate creates a dialect unique to Trinidad and Tobago.

Except in formal speech, the phoneme "th" is rarely used, with "d" or "t" being substituted, as in "Dis is jus' de t'ing". Trinbagonians often use nominative pronouns rather than objective or possessive pronouns and use present tense instead of past tense, as in "Yesterday he go to he house to wash he clothes". Trinbagonian expressions regarding location, weather, and time are constructed as they are in Romance languages rather than in formal English, such as "It have a man at the door".

aloo	white potato
Bacchanal	riotous party, usually involving alcohol
back-back	to reverse
Bad John	a dangerous man or troublemaker
BeeWee	British West Indian Airways (BWIA), the national airline of Trinidad & Tobago
Bhaigan	aubergine or eggplant
buss-up-shut	a kind of roti; from 'busted up shirt'
callaloo	a thick, green soup made from dasheen, okra, coconut milk, crab, and spices
calypso	a genre of music, usually with politics
calypsonian	a person who sings calypsos
catch a drop	hitchhike
channa	chick peas, garbanzos
clean	to have photographic film developed ('I need to get my flim clean').
crapo	(from the French *crapaud*, frog) a huge toad *(Bufo marinus)*
doubles	a biscuit-like East Indian staple food made with flour and crushed chick peas (channa)
fête	a social gathering with loud music, food, and dancing that usually lasts until well after midnight
fig	banana
flim	photographic film, a movie, or a video
galvanize	widely used zinc-coated metal sheeting used as roofing material
Good night!	unlike elsewhere, used in T&T in the evening in place of 'Good evening!'
Jumbie	an unfriendly spirit or ghost
jus' now	'hold on' or an indefinite time soon or later (often much, much later)
lime, liming	hanging around, talking, drinking, doing nothing in particular

mad	insane
melongen	aubergine or eggplant
pan	music played by a steel band using steel drums constructed from oil drums
pawpaw	papaya
pelau	a tasty dish made of rice, peas (or lentils or pigeon peas), and meat
petit carème	a dry period usually lasting the month of September
pine	pineapple
quenk	Collared Peccary, 'wild pig'
reverse back	same as 'back-back'
roti	a very filling East Indian baked dish made of flat dough wrapped around curried beef, chicken, conch, goat, shrimp, or other meat with onions, potatoes, tomato, and lots of spice
scrunting	lacking cash, as in 'Ramy is scruntin' now he job done.'
Soca	a genre of dance music with lyrics satirising politics and scandals, usually loaded with sexual content
Tattoo	armadillo
Vex	angry
Whappen?	'How's it going?' or 'What's going on?'
Zaboca	avocado
Zandolee	(from the French *les anoles*, lizards) any kind of gecko, lizard, or skink

SEASONAL DISTRIBUTION OF THE BIRDS OF TRINIDAD AND TOBAGO

The following section provides a quick means of determining the likelihood of sighting a particular species of bird at a specific time of year. No attempt has been made in this section to distinguish between relative abundances in Trinidad and in Tobago. Whatever information is available on the topic is contained in the section entitled "Selected Bird List" (see page 99).
The chance of finding a particular species is based on the assumption that the birder investigates appropriate habitat during the correct season and exercises due caution, avoiding loud noises, sudden movements, colourful clothing, and behaviour inappropriate to the quest.

KEY

(thick bar) Common – almost certain to be found every day in appropriate habitat and during the proper season.

(medium bar) Uncommon – possible to find with diligent searching in the appropriate habitat and during the proper season, but not to be expected every day.

(thin bar) Rare – not found on a regular basis. Present in small numbers, in limited habitat, or for only a brief period of time during the year.

.... Occasional/Accidental – species that keen birders long to see. Never to be expected; found only with great preseverance and more than a modicum of good luck.

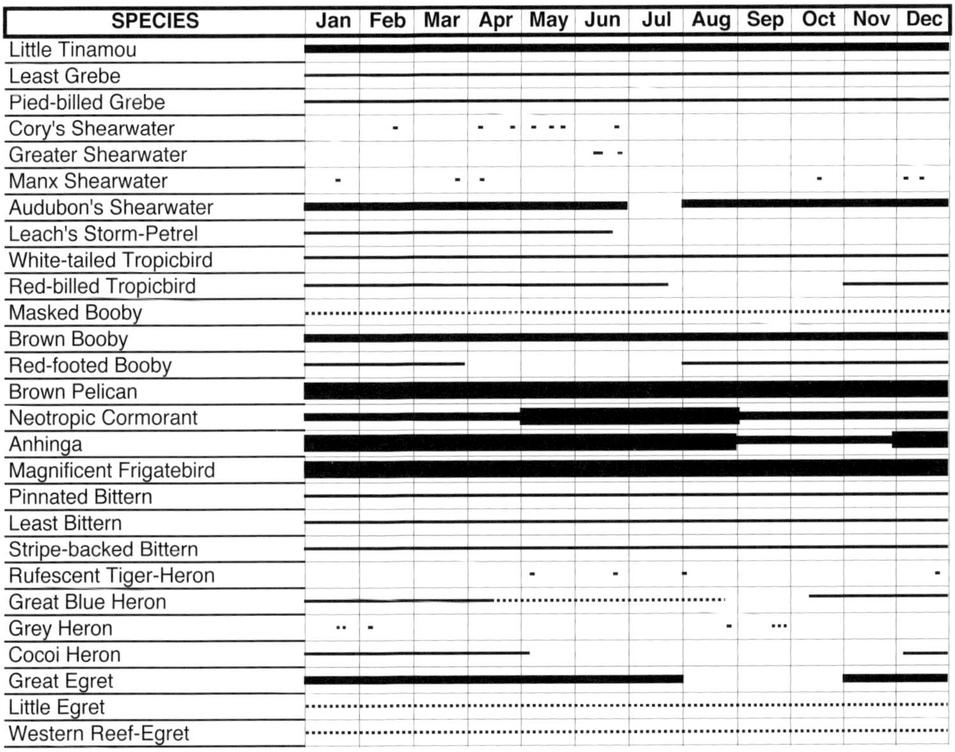

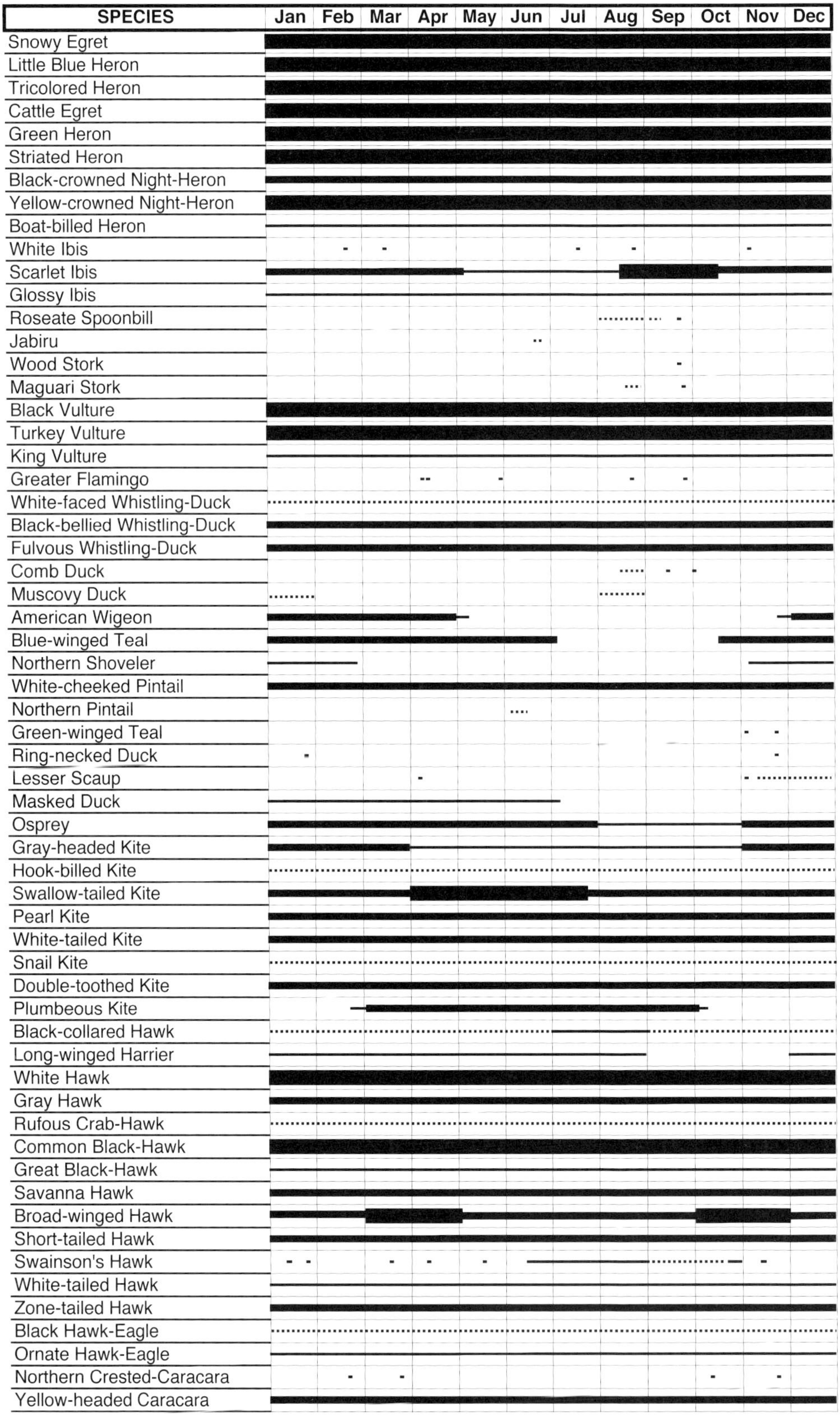
SPECIES
Jan
Feb
Mar
Apr
May
Jun
Jul
Aug
Sep
Oct
Nov
Dec
Snowy Egret
Little Blue Heron
Tricolored Heron
Cattle Egret
Green Heron
Striated Heron
Black-crowned Night-Heron
Yellow-crowned Night-Heron
Boat-billed Heron
White Ibis
Scarlet Ibis
Glossy Ibis
Roseate Spoonbill
Jabiru
Wood Stork
Maguari Stork
Black Vulture
Turkey Vulture
King Vulture
Greater Flamingo
White-faced Whistling-Duck
Black-bellied Whistling-Duck
Fulvous Whistling-Duck
Comb Duck
Muscovy Duck
American Wigeon
Blue-winged Teal
Northern Shoveler
White-cheeked Pintail
Northern Pintail
Green-winged Teal
Ring-necked Duck
Lesser Scaup
Masked Duck
Osprey
Gray-headed Kite
Hook-billed Kite
Swallow-tailed Kite
Pearl Kite
White-tailed Kite
Snail Kite
Double-toothed Kite
Plumbeous Kite
Black-collared Hawk
Long-winged Harrier
White Hawk
Gray Hawk
Rufous Crab-Hawk
Common Black-Hawk
Great Black-Hawk
Savanna Hawk
Broad-winged Hawk
Short-tailed Hawk
Swainson's Hawk
White-tailed Hawk
Zone-tailed Hawk
Black Hawk-Eagle
Ornate Hawk-Eagle
Northern Crested-Caracara
Yellow-headed Caracara

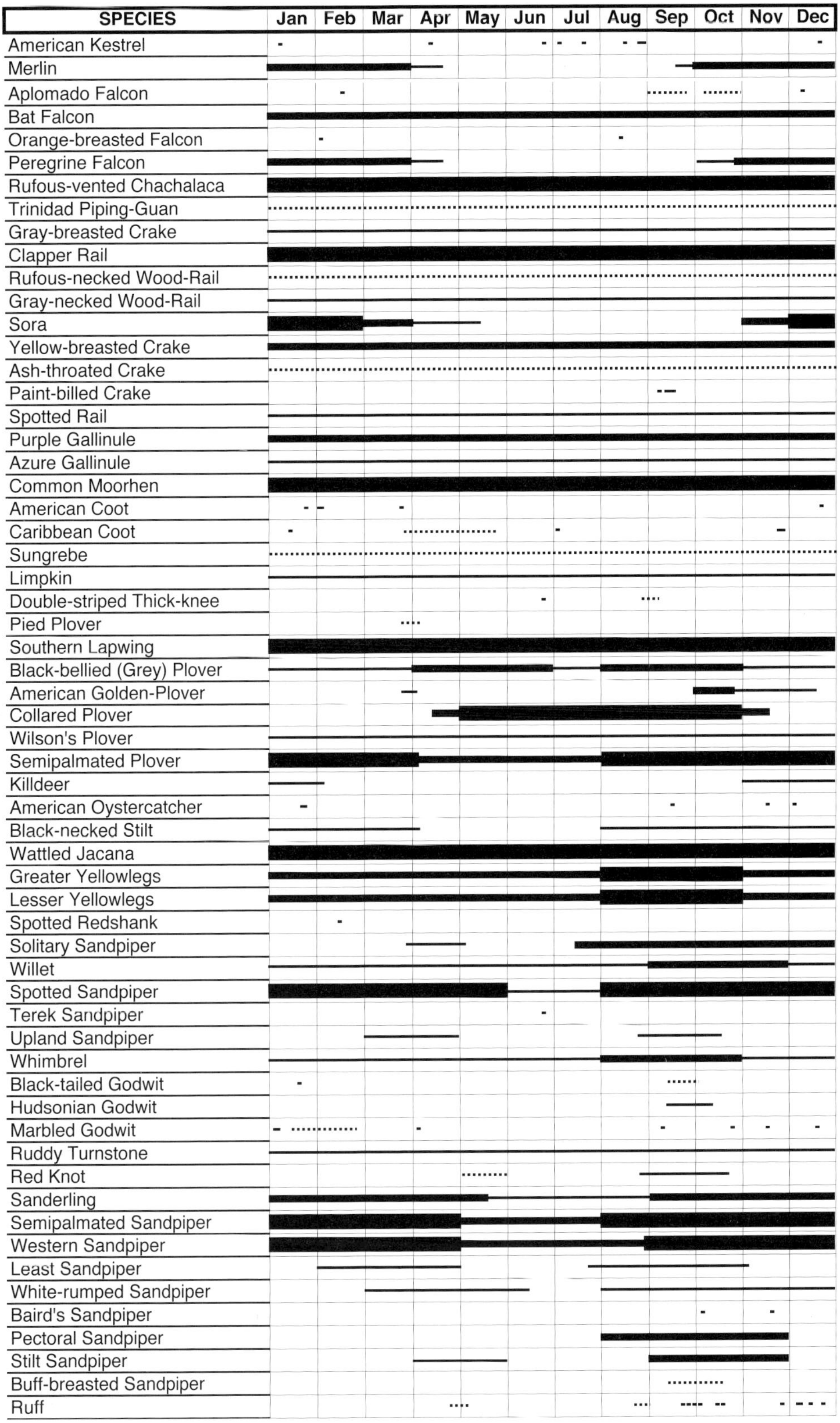
SPECIES
Jan
Feb
Mar
Apr
May
Jun
Jul
Aug
Sep
Oct
Nov
Dec
American Kestrel
Merlin
Aplomado Falcon
Bat Falcon
Orange-breasted Falcon
Peregrine Falcon
Rufous-vented Chachalaca
Trinidad Piping-Guan
Gray-breasted Crake
Clapper Rail
Rufous-necked Wood-Rail
Gray-necked Wood-Rail
Sora
Yellow-breasted Crake
Ash-throated Crake
Paint-billed Crake
Spotted Rail
Purple Gallinule
Azure Gallinule
Common Moorhen
American Coot
Caribbean Coot
Sungrebe
Limpkin
Double-striped Thick-knee
Pied Plover
Southern Lapwing
Black-bellied (Grey) Plover
American Golden-Plover
Collared Plover
Wilson's Plover
Semipalmated Plover
Killdeer
American Oystercatcher
Black-necked Stilt
Wattled Jacana
Greater Yellowlegs
Lesser Yellowlegs
Spotted Redshank
Solitary Sandpiper
Willet
Spotted Sandpiper
Terek Sandpiper
Upland Sandpiper
Whimbrel
Black-tailed Godwit
Hudsonian Godwit
Marbled Godwit
Ruddy Turnstone
Red Knot
Sanderling
Semipalmated Sandpiper
Western Sandpiper
Least Sandpiper
White-rumped Sandpiper
Baird's Sandpiper
Pectoral Sandpiper
Stilt Sandpiper
Buff-breasted Sandpiper
Ruff

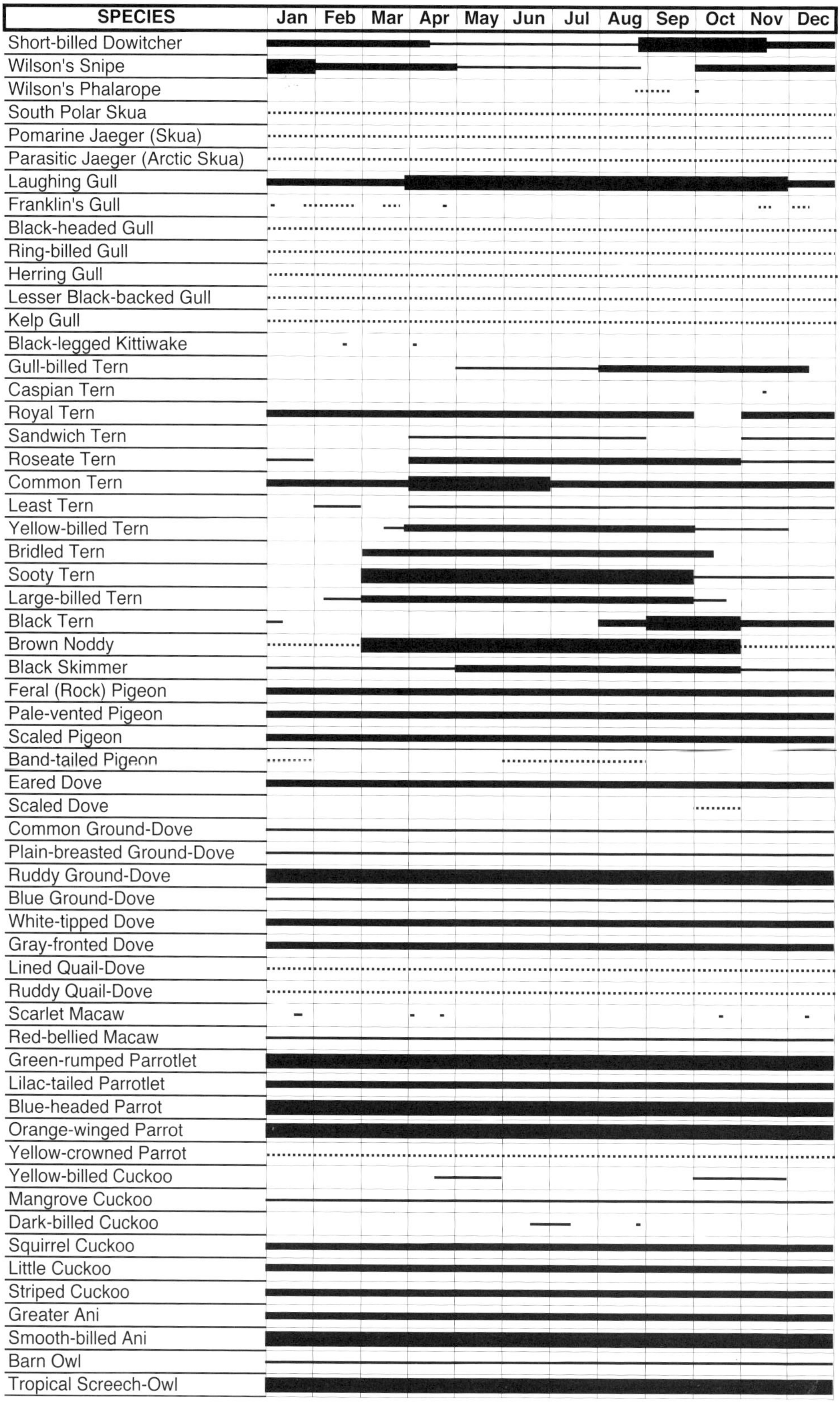
SPECIES
Jan
Feb
Mar
Apr
May
Jun
Jul
Aug
Sep
Oct
Nov
Dec
Short-billed Dowitcher
Wilson's Snipe
Wilson's Phalarope
South Polar Skua
Pomarine Jaeger (Skua)
Parasitic Jaeger (Arctic Skua)
Laughing Gull
Franklin's Gull
Black-headed Gull
Ring-billed Gull
Herring Gull
Lesser Black-backed Gull
Kelp Gull
Black-legged Kittiwake
Gull-billed Tern
Caspian Tern
Royal Tern
Sandwich Tern
Roseate Tern
Common Tern
Least Tern
Yellow-billed Tern
Bridled Tern
Sooty Tern
Large-billed Tern
Black Tern
Brown Noddy
Black Skimmer
Feral (Rock) Pigeon
Pale-vented Pigeon
Scaled Pigeon
Band-tailed Pigeon
Eared Dove
Scaled Dove
Common Ground-Dove
Plain-breasted Ground-Dove
Ruddy Ground-Dove
Blue Ground-Dove
White-tipped Dove
Gray-fronted Dove
Lined Quail-Dove
Ruddy Quail-Dove
Scarlet Macaw
Red-bellied Macaw
Green-rumped Parrotlet
Lilac-tailed Parrotlet
Blue-headed Parrot
Orange-winged Parrot
Yellow-crowned Parrot
Yellow-billed Cuckoo
Mangrove Cuckoo
Dark-billed Cuckoo
Squirrel Cuckoo
Little Cuckoo
Striped Cuckoo
Greater Ani
Smooth-billed Ani
Barn Owl
Tropical Screech-Owl

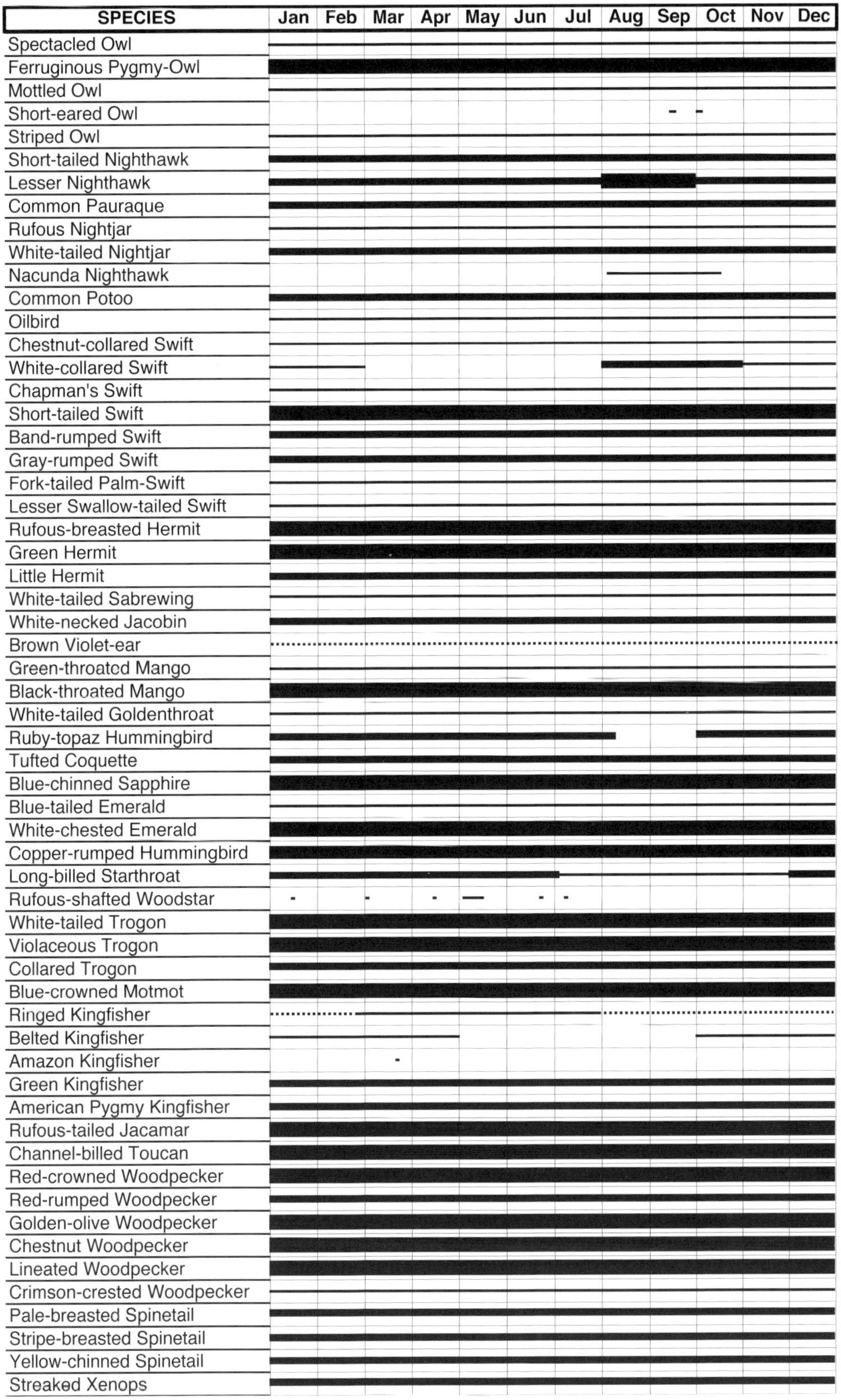
SPECIES
Jan
Feb
Mar
Apr
May
Jun
Jul
Aug
Sep
Oct
Nov
Dec
Spectacled Owl
Ferruginous Pygmy-Owl
Mottled Owl
Short-eared Owl
Striped Owl
Short-tailed Nighthawk
Lesser Nighthawk
Common Pauraque
Rufous Nightjar
White-tailed Nightjar
Nacunda Nighthawk
Common Potoo
Oilbird
Chestnut-collared Swift
White-collared Swift
Chapman's Swift
Short-tailed Swift
Band-rumped Swift
Gray-rumped Swift
Fork-tailed Palm-Swift
Lesser Swallow-tailed Swift
Rufous-breasted Hermit
Green Hermit
Little Hermit
White-tailed Sabrewing
White-necked Jacobin
Brown Violet-ear
Green-throated Mango
Black-throated Mango
White-tailed Goldenthroat
Ruby-topaz Hummingbird
Tufted Coquette
Blue-chinned Sapphire
Blue-tailed Emerald
White-chested Emerald
Copper-rumped Hummingbird
Long-billed Starthroat
Rufous-shafted Woodstar
White-tailed Trogon
Violaceous Trogon
Collared Trogon
Blue-crowned Motmot
Ringed Kingfisher
Belted Kingfisher
Amazon Kingfisher
Green Kingfisher
American Pygmy Kingfisher
Rufous-tailed Jacamar
Channel-billed Toucan
Red-crowned Woodpecker
Red-rumped Woodpecker
Golden-olive Woodpecker
Chestnut Woodpecker
Lineated Woodpecker
Crimson-crested Woodpecker
Pale-breasted Spinetail
Stripe-breasted Spinetail
Yellow-chinned Spinetail
Streaked Xenops

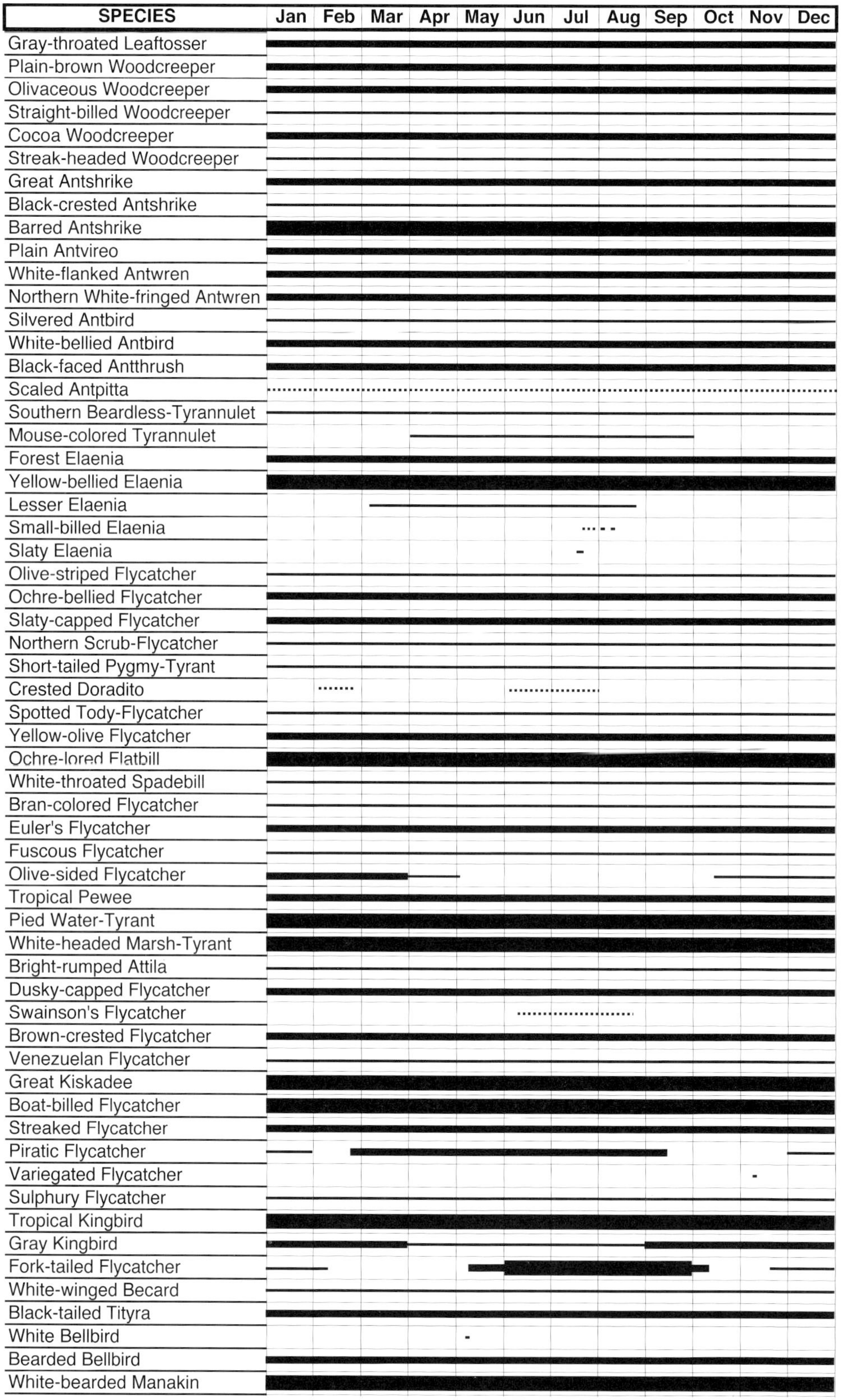
SPECIES
Jan
Feb
Mar
Apr
May
Jun
Jul
Aug
Sep
Oct
Nov
Dec
Gray-throated Leaftosser
Plain-brown Woodcreeper
Olivaceous Woodcreeper
Straight-billed Woodcreeper
Cocoa Woodcreeper
Streak-headed Woodcreeper
Great Antshrike
Black-crested Antshrike
Barred Antshrike
Plain Antvireo
White-flanked Antwren
Northern White-fringed Antwren
Silvered Antbird
White-bellied Antbird
Black-faced Antthrush
Scaled Antpitta
Southern Beardless-Tyrannulet
Mouse-colored Tyrannulet
Forest Elaenia
Yellow-bellied Elaenia
Lesser Elaenia
Small-billed Elaenia
Slaty Elaenia
Olive-striped Flycatcher
Ochre-bellied Flycatcher
Slaty-capped Flycatcher
Northern Scrub-Flycatcher
Short-tailed Pygmy-Tyrant
Crested Doradito
Spotted Tody-Flycatcher
Yellow-olive Flycatcher
Ochre-lored Flatbill
White-throated Spadebill
Bran-colored Flycatcher
Euler's Flycatcher
Fuscous Flycatcher
Olive-sided Flycatcher
Tropical Pewee
Pied Water-Tyrant
White-headed Marsh-Tyrant
Bright-rumped Attila
Dusky-capped Flycatcher
Swainson's Flycatcher
Brown-crested Flycatcher
Venezuelan Flycatcher
Great Kiskadee
Boat-billed Flycatcher
Streaked Flycatcher
Piratic Flycatcher
Variegated Flycatcher
Sulphury Flycatcher
Tropical Kingbird
Gray Kingbird
Fork-tailed Flycatcher
White-winged Becard
Black-tailed Tityra
White Bellbird
Bearded Bellbird
White-bearded Manakin